W9-AQT-626

A Logical Theory of Teaching
Erotetics and Intentionality

Philosophy and Education

A Logical Theory of Teaching

Erotetics and Intentionality

by

C. J. B. MACMILLAN
The Florida State University

and

JAMES W. GARRISON
Virginia Polytechnic Institute and State University

KLUWER ACADEMIC PUBLISHERS
DORDRECHT / BOSTON / LONDON

Library of Congress Cataloging in Publication Data

Macmillan, C. J. B. (Charles James Barr), 1935-
A logical theory of teaching : erotetics and intentionality / by C.J.B. Macmillan and James W. Garrison.
p. cm. -- (Philosophy and education)
Bibliography: p.
Includes indexes.
ISBN 9027728135
1. Questioning. 2. Teaching. 3. Intentionality (Philosophy)
I. Garrison, James W., 1949- . II. Title. III. Title: Erotetics and intentionality. IV. Series.
LB1027.44.M33 1988
371.1'02--dc19 88-13638
CIP

ISBN 90-277-2813-5

Published by Kluwer Academic Publishers,
P.O. Box 17, 3300 AA Dordrecht, The Netherlands.

Kluwer Academic Publishers incorporates
the publishing programmes of
D. Reidel, Martinus Nijhoff, Dr W. Junk and MTP Press.

Sold and distributed in the U.S.A. and Canada
by Kluwer Academic Publishers,
101 Philip Drive, Norwell, MA 02061, U.S.A.

In all other countries, sold and distributed
by Kluwer Academic Publishers Group,
P.O. Box 322, 3300 AH Dordrecht, The Netherlands.

Printed in The Netherlands

TABLE OF CONTENTS

ACKNOWLEDGMENTS

We owe more than we can express to many people, journals and government institutions. Among the people are most importantly Jaakko Hintikka, Shirley Pendlebury, Robert H. Ennis, David P. Ericson, and Frederick Ellett, Jr. Hintikka was seminal in our thinking, as the most cursory reading of what follows will show. The others have responded - not always with compliments - to various earlier versions of the ideas presented here.[1] James T. Dillon gave moral support to this study, and provided us with information about questioning that is available from no other source. Kenneth Strike and Denis Phillips have read earlier drafts of the complete volume; their comments have prevented us from making some serious errors. All of these colleagues, of course, are absolved of responsibility for our confusions and errors. We have not always taken their advice.

The journal *Educational Theory* was the place where most of our ideas were first played out in public. It began with Macmillan and Garrison (1983), continued with the critique of process-product research (Garrison and Macmillan 1984), and continued with responses to critics in later years (Macmillan and Garrison 1986, 1987). Close readers of this volume will recognize many

1 See Pendlebury (1986) and Ennis (1986) for very important comments; to some extent we have tried to bring our response (Macmillan and Garrison 1986) into this volume, particularly in Chapter II. Ericson and Ellett (1987) took a different tack, examining our ideas about causation very closely. Again, we have brought part of our response (Macmillan and Garrison 1987) into the body of this work in various places. These critics were always helpful to us - because of their close and sympathetic reading, we hope that we have been clearer than otherwise would have been possible.

passages that appeared there. Most especially, Chapter II, which includes most of the original paper "An Erotetic Concept of Teaching" (1983) and a fair amount of "Erotetics Revisited" (1986). Chapter III is a slightly revised version of Garrison and Macmillan (1984). Chapter X includes still more of the 1983 paper. All of these are reprinted with the kind permission of the Editor of *Educational Theory*. To that journal and its editor, Ralph Page, we owe great thanks.

A portion of Chapter I appears in *The Journal of Research and Development in Education* (Garrison and Macmillan 1987). It is reprinted here in revised form with the permission of the editor, George Newsome.

Chapter VIII was originally part of an interdisciplinary study of questions in education headed up by James Dillon. It is a revised version of a chapter appearing in Dillon (1988). It - and the transcripts of classrooms - are reproduced with the permission of the publisher, Ablex, and of Dillon himself.

The diagram on page 68 is reprinted by permission of the publisher from Gage, N. L., THE SCIENTIFIC BASIS OF THE ART OF TEACHING. (New York: Teachers College Press, c 1978 by Teachers College, Columbia University. All rights reserved.), p. 73.

The quotations and the diagrams from Plato's *Meno* in Chapter VII are reproduced by permission of Penguin Books Ltd. from *Plato: Protagoras and Meno,* Translated by W. K. C. Guthrie. (London: Penguin Classics, 1956), copyright (c) W. K. C. Guthrie, 1956, pp. 130–138.

Support for the research that went into this volume was provided by the National Science Foundation, which supported a grant entitled "Questioning as a Knowledge-Seeking Method;" principal investigators were Jaakko Hintikka and C. J. B. Macmillan. Without that support, this work would have been impossible.

Professor Alan Mabe, of Florida State University's Department of Philosophy, oversaw the typesetting of this book. His time and effort deserve more gratitude than we have to offer.

Florene Ball, of the Department of Philosophy at Florida State University, took us and the word processor through many drafts, always with speed, accuracy and good humor. To her go more thanks than we can express here or in person.

We dedicate the volume to our wives, Joan Macmillan and Leigh Garrison; they provided the sort of moral - and even editorial - support that can only be gotten from such close friends.

CHAPTER I

INTRODUCTION: THE INTENTIONALIST MANIFESTO

Over the past thirty years or so philosophers of education have made repeated assaults on the dominant tradition of educational research. This research tradition can be characterized, perhaps too roughly, as behavioristic in theory, statistical in methodology, and positivistic in epistemology and metaphysics. It has reared its head in almost every area of educational thinking and has ridden many waves of political and administrative faddism despite the outraged cries of philosophers from Dewey (1916) to Scheffler (1960) to Green (1971). But philosophers' arguments have had remarkably little effect; in part this may be the result of - and evidence for - the relatively weak place that philosophically oriented people have in the educational establishment.

In large part, though, we suspect that the lack of effect is because the philosophers' cries have seemed beside the point. The response of a Skinner, a Gage, or whoever, has been, "Isn't it nice that we can afford to worry about meanings and values? But let's get back to work, now: What methods best cause children to learn what we want them to learn?" With that, they return to analyses of variables, meta-analyses of empirical studies, and various forms of experimentation that depend on a model that philosophers have tried to argue is wrong-headed from the start.

We are in new times, though; in the past twenty years, there has been a break in philosophy of science, a change from pro-science positivism - whose enemies tended to be anti-scientific philosophers of various sorts - to the equally pro-science but broader vision of post-positivistic philosophy of science. (For a discussion, see Garrison, 1986.) Probably *the* lesson to be learned from this break is that no scientific theory and no larger context of scientific theories (be it called 'paradigm', 'programme', or 'tradition') can be fruitfully assessed apart from competing theories or traditions.

And this, we fear, is where philosophers of education have managed to miss their chance. Excellent at criticism, excellent even at morality and metaphysics, philosophers have not been so good at providing real options for those who believe that education can and should be studied scientifically. The challenges have been missed in large part because they often seem to suggest that the scientific study of educational matters is either impossible or irrelevant. (See for example, Tom 1980, or the NSSE Yearbook entitled *The Philosophical Redirection of Educational Research* (Thomas 1972), notable as much for the arrogance of its title as the profundity of its contents.)

Doomsaying seems relatively easy to philosophers; we try to show that something is logically impossible - or at least invalid - but empiricists just keep on doing it. Like bumblebees, they fly from flower to flower gathering in the pollen for their own purposes despite our claims of impossibility. Perhaps philosophers have missed the point.

Let us instead accept one basic assumption of the empiricists: it is possible to study education empirically, using highly sophisticated techniques and methods of investigation.

In fact, let us *welcome* this assumption, for to reject it leads eventually to an abhorrent conclusion - that there can be no empirical study of education, that in effect the attempt to learn from experience is doomed from the start. At bottom, after all, a scientific approach is little more than a set of variations on the belief that we *can* learn from our own and others' experience, learn what happens, how it happens, and why it happens. Our assumption ought to be that this is as true in education as it is in atomic physics.

But this leaves many other questions to answer. The crucial ones: What *kind* of science is proper or appropriate to education? How does it *differ* from physics? What is wrong with the prevailing, virtually unopposed research tradition in education? What could or should be done to replace it with a more adequate tradition? What concepts are necessary to describe and explain what we find there? It is in this realm that we find ourselves.

Where to start? One place - our place, needless to say - is with one limited but central concept in education, *teaching*. A long philosophical tradition concerned with the nature of teaching goes back (along with everything else) to Plato, divulging most recently in the work of such philosophers as B. O. Smith, Scheffler, Hirst, Komisar, Green, McClellan, Soltis, Kerr, Fenstermacher, *et al.* An empirical tradition runs parallel to the philosophers - it has its most notable modern proponents in Gage, the Soars, Berliner, Rosenshine, but its roots can be traced to the Sophists. These two traditions have been at loggerheads over the centuries.

One line of philosophical criticism occurs again and again, met most often by blank stares or cries of "*Unscientific!*" from the research community. This is that teaching in particular and human action in general is irreducibly *intentional*, that human beings act in accord with *beliefs* that they hold about the *meaning* of things in their environment, for the achievement of *goods* or *values*.

The dominant modern tradition of research on teaching and learning has tried to get around the features of belief, meaning and value that are central in intentionalistic approaches. Where these have not been outright rejected as of no scientific interest, they have been ignored because they were or are not capable of treatment in the "rigorous" statistical analyses that are the methodological foundation of that tradition. Philosophical potshots have been of little avail, for philosophers have not shown just how there can be a rigorous empirical approach to intentionality in education and teaching. Even the present war between quantitative and qualitative research methods - which gets at this issue - often boils down to a contrast between a hard-headed realistic epistemology and ontology and a viciously relativistic idealism. (See, for example, J. K. Smith 1983.)

Our suggestion is this: try to develop an approach to teaching which is both empirically rigorous and intentional. One way into this is through the philosophical tradition represented most notably by Komisar (1968), T. F. Green (1971), and McClellan (1976), in their analyses of the concept of teaching. Show how this approach can deal with the same questions raised by the empirical tradition without falling into the anomalies that the latter is heir to. Show how much more can be done when one attends to what is ordinarily recognized as teaching, without having to go through a series of translations from research to practice and back again.

The "intentional" features of teaching that such an approach has to deal with include the following points. Teaching is intentional in two senses. First, the teaching itself must be seen as intentional action; minimally, this means that showing that one did not intend to teach something to someone else is grounds for saying that one did not teach it, even though the learner might have learned it because of something the "teacher" did. "Why did you teach your grandson to eat peas with his knife?" we query the old man. "I didn't teach him that," he responds, "he just copied my way of eating." And we can accept his plea - at least in part.

The second relevant sense of intentionality is in the logical and epistemological realm - where "propositional attitudes" enter into logic, where the "aboutness" of mental states enters epistemology and philosophy of mind, or where the semi-grammatical nature of "intentional objects" is in the foreground. (See Quine 1966, Searle 1983, and Anscombe 1965, for the different approaches encapsulated by these different characterizations.) Minimally, one teaches *something* to someone else, and that something is intentional in the sense of being a belief, a desire, or perhaps a way of acting, given beliefs and desires. Teaching is double-barrelled on this score, though, since there is always another person involved in the interaction - the learner (or the student, to be less deterministic about it). The learner's relationship to the teacher is intentional in this second sense at least and arguably in the first sense as well. (Hirst, 1971, argues that "learning" should be viewed as an intentional action.)

Note that we are avoiding what may be the most significant philosophical issue to be found here - whether intentionality in either of these two senses is to be understood in terms of the other. This problem is intrinsically interesting and worth attention; in this context, however, it is enough to recognize that there are two aspects to intentionality; settling this issue will have to await the future. (See Anscombe 1965 and Searle 1983 for opposing positions.)

To translate this into action, into the tradition of educational research and practice, we propose the following intentionalist manifesto.

An Intentionalist Manifesto for Research on Teaching

Article 1: People *believe* things about the world. Belief is a patently intentional notion; it is, however, non-veridical. Belief and true belief are distinct, and both of them are distinct from knowledge, although all are related in pedagogically important ways. The commonsensical assertion that people believe things about the world is far from being noncontroversial. Many philosophers, theoreticians and researchers look upon beliefs as creatures of darkness, a constant threat to the radiant sunshine of objectivity; vague, inexact, ill defined and subjective at best, chimerical and false at worst.

Article 2: People's beliefs change. This we take to be one of the most dramatic and least debatable conclusions (or assumptions) of cognitive psychology, not to mention the whole process of education. Something changes, that seems apparent. We, at least, do not hesitate to call that something beliefs.

Article 3: Changes in beliefs can be explained. What is more, the explanations may be rational, even logical, although the pattern of explanation may differ from the traditional structure of explanation typified by the covering law model.

Article 4: The explanations of these changes in belief are (at least sometimes) causal explanations.

Article 5: Teaching must involve attempts to change beliefs; and, insofar as it does, it must include a causal analysis of some sort.

Article 6: Teaching may function as a cause of changes in belief. This does not exclude other explanations about what takes place in learning, but only suggests some limitations and attempts to overcome them. We hope to show one (logical) way this may be done.

Article 7: Teaching acts, functioning as rational causes of changes in beliefs, must involve the *meaning* (semantic content) of the beliefs.

Article 8: The cause of a belief is, at least sometimes, also a justification for the belief. Why not always? Consider two forms of explanation of beliefs: "I believe *p* because my parents did," vs "I believe *p* because of evidence for *p*." Both involve intentional causes and both involve reasons, but only the latter implies a rational justification for the belief. In any case both must be honored by researchers. There is, perhaps, a second reason why the cause of a belief is only sometimes its justification. Consider the case, deployed by MacIntyre (1967) against an exclusively interpretivist view of social science. MacIntyre bids us consider an act (or, presumably, a belief as well) imposed by post-hypnotic suggestion. Here the explanation appeals to a cause that is largely if not entirely physicalistic. It seems to MacIntyre, and to us, that such an explanation may not appeal to reasons at all, much less justifications, as the cause of a belief. This is an important point.

Intentional research on teaching need not conceive itself as competing with more traditional forms of research on teaching. Rather, intentional research may be seen as supplementing traditional approaches in areas where the methods of the latter are problematically used.

Article 9: Teaching, at its logical or conceptual core, involves rational causes.

Article 10: Teaching acts and the belief contexts they give rise to may be studied in a manner that is as logically rigorous and precise as any to be found within the citadel of positivism. Note we say *may* be. We believe, as a rule, they should be; but one may well adhere to the intentionalist manifesto for research on teaching without accepting this principle also.

In what follows we attempt to reconceptualize research on teaching in accordance with this intentionalist manifesto. Our goal is to provide an intentionalist *theory* of teaching, one that embraces the criteria of precision and rigor indicated by Article 10. But the attempt to construct a theory of teaching prior to empirical research on teaching runs afoul of another widely accepted empiricist dogma. The answer to this dogma is not another manifesto but a plea; a plea for theory in research on teaching.

Educational Research to Pedagogical Practice: A Plea for Theory

The problem for any practical or professional field is how to put the results of research into practical use. In Aristotelian terms, how does theoretical knowledge become practical knowledge? It was this problem, we suspect, that led to the beginnings of the current research tradition in teaching, for no pedagogical practice follows directly from the fact that learning curves for lists of nonsense syllables are J-shaped. No teacher can find guidance in this "fact" alone to help in pedagogical planning and practice. But the attempt to by-pass this problem by attending directly to teaching only makes it come up in a different guise. How do such findings as that the correlation between (a) time-spent-on-learning-task and (b) student achievement falls into an "inverted-U" curve (Soar and Soar 1976) relate to teaching any particular subject to particular children? Again, no particular practical directives can be drawn from the research finding alone; much else is needed.

This leads to a search for ways of relating research to practice. Many writers have examined the issue recently (Fenstermacher 1979, Tom 1980, 1985, Ericson and Ellett 1982, Macmillan and Garrison 1983, for example). We shall not here review all the arguments and problems they have brought up. It is enough to note that all agree that the conversion of research findings into practical knowledge is possible only if we understand the conversion process more adequately.

One place to look for a clue to understanding the conversion process is to a successful profession; the most obvious parallel is to be found in medicine, where the profession can almost date its practical successes to the relatively recent emergence of a research base adequate enough to support effective practice. It is interesting to note that this research base did not begin to emerge until researchers turned their immediate attention from medical practice narrowly conceived, to biomedical research, specifically bacteriology. There are no exact dates for the emergence of a research base sufficient for the theoretical support of medical practice, although 1928, the year Alexander Fleming discovered penicillin, is certainly a landmark. But by then the German bacteriologist Paul Ehrlich had been seeking chemicals that could serve as "magic bullets" to destroy microorganisms for nearly 30 years, and it would be another 13 years before penicillin could be mass produced and have a wide medical application (Wilson 1976). Optimists like Gage (1978, p. 92) and ourselves draw inspiration from the history of medicine and are hopeful that the shift of attention from teaching practice, and even learning research, to research on teaching may have a similarly happy result.

In two recent papers Gage (1983, 1984) uses a medical analogy. He does not compare educational and medical practice (an analogy fraught with problems), nor even educational and medical research. Rather he compares the use of research findings in the two fields; particularly, he considers the use of comparable statistical results as the basis for practical recommendations. We want to examine this comparison fairly closely, for we think that facile acceptance of the analogy carries perils that are not obvious on the surface.

The first break in the analogy is this: Medical researchers make precisely the same assumptions about the world as are made by medical practitioners. On both sides of the research-practice gap, the language, logic, and evidence are the same; the causes of biological and medical conditions and changes are seen as the same by researchers and practitioners and the conversion from research to practice is almost direct. In order to achieve a reduction in pain, give aspirin, because research has found that aspirin brings about pain-reduction.

This is not clearly the case with current research on teaching and pedagogical practice. The former proceeds by assuming an explicitly *non-intentional* model of the relationship of teacher and student, by quantifying the "behaviors" observed, i.e., by attending only to those aspects of the teacher-student-subject matter relationship that can be measured and correlated according to statistical procedures which had been developed in contexts like agricultural and medical research. (It is interesting to note that in 1963, Gage suggested an explicitly agricultural model for educational theory, while in 1984, he appeals to medical statistical procedures.) But in practice, teachers must speak specifically about particular subjects and students. And this requires treating the subjects taught as having a particular *meaning* and the students' achievement of the goals of teaching in the same (meaning) terms. The problem of conversion of research findings here is also a problem of *translation* from the explicitly non-intentional language of the researcher to the explicitly intentional language of the pedagogical practitioner.

This translation is multi-layered, involving as it does an institutional context (with the implicit values and goals put upon schools by our social, political and economic systems), the teachers' intentions (which may or may not gibe with the schools'), and the students' intentions (ditto). More important for research and practice, however, is the fact that the relationship of teacher, student and subject matter is at the lowest level filled with meaning in ways that the doctor-patient relationship is not; when a doctor says, "Take two aspirin (or a beta-blocker) and call me in the morning," the patient can understand and follow or refuse to follow the recommended treatment. But the teacher's statement, "Washington crossed the Delaware at midnight on the 25th of December," *is* the pedagogical "treatment". Current research on teaching avoids dealing with this basic feature of pedagogical practice. Medical research says that aspirin is effective and attempts to say why; it does not need intentions or intentionality to answer the why-question. Educational research requires intentionality to explain its findings about teacher effectiveness. This incommensurability cannot be bridged, as Gage (1978) at one point attempts, by simply turning the results of teaching research over to the practical artists - the teachers - for they, wise only in the ways of intentionality, may not understand how to use the results, or worse still, may misuse them. More on this point later.

Gage suggests that recommendations can be made for educational practice based upon statistical results that would be adequate for recommendations in medical contexts. This is the site of the second break in the medicine/education analogy. Gage compares an actual research project in medicine to an imaginary experiment in education. In the former (a beta-blocker experiment), medical researchers were willing to recommend treatment based upon statistical results that would not be used in educational practice; in education, Gage states, the same figures would imply "hardly a 'shred'" of a recommendation (1983, p. 494).

Gage's diagnosis of the difference is that it is merely a matter of cost-benefit relations: "When the cost is low and the benefit is great, even a weak 'main effect' has powerful implications" (1983, p. 494). But Gage overlooks a crucial feature of the medical example. The beta-blocker experiment took place within a theoretical context that is widely accepted by medical researchers and practitioners alike. When one finds a lack of an accepted theoretical context, the same statistical correlations do not lead to such sharp agreement over recommendations for medical practice, even where the costs are low. In a study of the diets of hypertensives, for example, it was found that hypertensives had significantly lower calcium intake than did non-hypertensives but that their sodium intake was not significantly different. "After controlling for age, race, and sex, difference in dairy product consumption proved to be the best predictor of hypertension" (McCarron et al, 1984, p. 1394). Based upon these findings, the authors were willing to make cautious recommendations, for example that calcium intake be carefully monitored when low-sodium diets are prescribed for hypertension (1984, p. 1297). But the study raised a storm of controversy among commentators. (See Kolata 1984, for a summary.) While McCarron is willing to make recommendations, others argue that the data on calcium are "not *nearly* at the level to make dietary recommendation" (Friedewald, quoted by Kolata, p. 706). Another set of critics attacks the findings on theoretical (conceptual) and methodological (statistical) grounds rather than recommending action (Feinleib, Lenfant and Miller 1984).

The moral of this story is that medical recommendations are made only where there is wider agreement on the theoretical background for the statistical findings; critics "ask what sort of physiological mechanism could possibly account for such calcium effects" (Kolata 1984, p. 705), for example, and where a theoretically acceptable account is not forthcoming, the medical community is extremely hesitant to make strong recommendations, however high the cost-benefit ratios. Thalidomide casts a long shadow in these contexts.

Pedagogical theory may not be a matter of instant life and death. But the logic of the change from pedagogical theory to practice deserves attention if educational practice is to have anything like the knowledge base of medical practice. This is too complex an issue to resolve here, but some features of the pedagogical situation deserve mention for those who would attack the question.

The point of research on teaching, presumably, is to provide guidance to teachers, whose knowledge of the pedagogical situation in which they find themselves would be broadened by studying research findings as a doctor would study the results of medical research. In some way - perhaps not directly (whatever *that* might mean) - the research should provide guidance for teachers. Teachers come to the pedagogical setting with a bundle of ideas about teaching, about its purposes and goals, about the best modes and methods of approaching certain subjects with specific students. We can grace this bundle of ideas with the title of "theory,"' as long as we recognize that for the usual teacher, these ideas are not organized in any specific way, that they are intuitively rather than objectively reasonable for them, and that they are based upon their own limited personal experience, perhaps poorly interpreted.

Research on teaching has to be filtered through these teachers if it is to have an effect on schools. This means that in some way it has to be used to broaden their own experience, to give an objective base to those personal theories, perhaps to break into the intuitions, to show there they are wrong and right and *why*. In a significant sense, the objective findings of research on teaching can be viewed as competing with the intuitive knowledge that teachers carry with them. The ideal might almost be viewed as the attempt to replace one paradigm or scientific tradition with another, to use the new way of thinking to explain what is right about the old and to show why the old way is unsatisfactory for the explanation of the world and the direction of practice. What is required for such learning to take place is an explicit clash of theories, as the individual subjective theories of teachers compete with objective research-based theories of teaching.

But the non-theoretical approach of most current research on teaching makes this clash impossible in principle. Gage states the case:

> At the least, a scientific basis [for the art of teaching] consists of scientifically developed knowledge about relationships between variables. . . .
>
> Notice that I say nothing about theory, nomological networks, systems of postulates and axioms, or hypothetico-deductive relationships. In emphasizing the relationships between variables, I am not denying the desirability of systematic theory; I am merely saying that, however desirable, systematic theory is not indispensable to any valid conception of science. The grand theories in the natural sciences and even theories of the middle range in the behavioral sciences may represent goals for us to aspire to, but they do not set minimum requirements as to what we must have before we can lay claim to scientific knowledge. (1984, p. 89)

But the question remains: Just what are the minimum requirements of scientific knowledge? Facts, Gage seems to suggest, general facts in the form of correlations between variables. But this won't do as a picture of science, or even perhaps of personal knowledge of the world; as one educational researcher (Anderson 1984) has put it, "knowledge is not a 'basket of facts'" (p. 5). We might add that a non-theoretical approach is worse still when practical uses of knowledge become the goal.

Let us explain what makes it impossible for the findings of such research to be a satisfactory guide to the ordinary teacher's practice.

Facts are promiscuous. They will consort with any number of interpretations and misinterpretations. Facts, immodestly coupled with other facts, all too frequently yield to an undesirable degree of theoretical excess. Any fact may be used to support or to disprove any theory that is at a more general level than the fact itself - depending, of course, on the interpretation we give it. "That tree is brown" is contradicted by its being any other color at the moment; but this sentence is not an outright falsification of the arborological generalization, "trees are green," since the qualifications and exceptions to the more general statement are built into its theoretical context. Only the theoretical context of the individual statement or generalization makes it significant for practical or explanatory use.

Merely to provide "facts" or "findings" of non-theoretically driven research is to provide teachers with no help in developing their own theories of the pedagogical situation. For the most likely outcome is that such facts will simply be tailored to fit the already existing intuitive theory; the significance *of* the facts is not given *by* the facts, so there is no guarantee - or even likelihood - that the findings of research will be of use except where they are put into an already existing theory. But what is necessary is to help teachers develop objectively defensible theories - "objectively reasonable beliefs" to use Green's (1971) and Fenstermacher's (1979) term - and the theoryless fact cannot do this alone.

But we haven't reached bedrock yet. Facts may confront theories, including subjective or intuitive theories, but they need not refute them. As Duhem (1906, 1956) and Quine (1951) have shown, a clever and determined advocate of even objectively reasonable theories can always defend a selected core of propositions on logically proper grounds, come what may in the way of falsifying examples. The result of confronting subjectively or intuitively reasonable theories with uninterpreted facts will be many intuitively reasonable theories supported by objective facts subjectively interpreted. This has been exemplified recently in a study by Kennedy (1984), who investigated many different episodes in which educators interacted and interpreted facts in order to arrive at practical decisions. Kennedy explored the "relativity of meaning," showing how the meaning of facts is dependent on the wider interpretive context including, among other things, the participants' background knowledge, beliefs, experiences, interest, and even values. This is closely related to what we refer to as the teachers', or more broadly speaking, educators', intuitive or subjective theories of the (pedagogical) world. When the researcher's findings are only put into theoretical contexts after the fact, no theories get tested - not the researchers' theories, not the teachers' theories, not anyone else's theories. It is only when the research is theory-driven from within that the theories themselves are supported or falsified; and only when this is done can there be better theories for the explanation and direction of pedagogical practice.

So this is the plea: don't give us mere facts or findings, however resounding the statistics. The Promethean gift of statistics may burn the hands of its most competent users. Give us theoretically driven research, where we may disagree or agree on the import of the theory as well as the facts. With this beginning, we can argue about the correct approach from an objectively defensible starting point. Otherwise there is just the blooming buzzing confusion of the uninterpreted world or the worse confusion of missed interpretations.

An Erotetic Theory

The theory we propose is based upon an erotetic concept of teaching. We analyze teaching as a question-answering activity and develop our theory from that analysis. On the erotetic concept it is the intention of teaching acts to answer questions that students ought to ask concerning the subject matter with which the teacher and student are engaged. Note that the defining characteristic of *teaching* in this context is not *learning* but rather a formal characteristic of the interaction between teacher and student. The fruitfulness of this turn away from learning is part of what we hope to demonstrate in the rest of this book. We think it important to have a theory of teaching that approaches research on teaching as a domain separate from other areas of inquiry in education - especially learning. At this juncture, however, we can only beg the reader's indulgence and hope to be heard to the end.

Ours is a formal and not a substantive theory of teaching; it may (and will) be codified by erotetic logic, i.e., the logic of questions and answers. Our specific choice from among several competing forms of erotetic logic, that of Hintikka and Åqvist, captures the kind of precise and rigorous formulation called for by Article 10 of the intentionalist manifesto.

The erotetic theory of teaching is not substantive in the sense that it does not *directly* say anything about *what* should be taught, about the aims of education, or even very much about *how* to teach, although what it does say carries the weight of logical necessity. It might be considered a weakness that a theory of teaching does not include answers to questions of aims and "methods" of teaching. Our austere theory of teaching, it would seem, cannot say much of interest to researchers or practitioners. We don't agree, of course: in our opinion many educational problems lie outside the domain of teaching. A theory of teaching ought to provide a framework for identifying and discussing those problems, but it alone cannot solve them. One of our tasks will be to show how our theory helps to clarify other areas of educational decision-making. In any event there are decided advantages of a formal theory such as ours. Let us consider a few of these.

The greatest advantage of our theory is that it provides very exact criteria for successful teaching. These criteria derive from logical, semantical and epistemological conditions for satisfactorily answering questions. Such clear and distinct criteria of successful teaching make it easy to put the erotetic theory to work in the planning, implementation and evaluation of teaching. A formal theory like this leaves a great deal of latitude for the individual teacher's methods, style and approach in planning and implementing a lesson and in the evaluation of that lesson, something currently popular theories and research paradigms of teaching do not always do. Since the erotetic theory is not specific about the content of teaching, its formal structure may be satisfied in any number of ways; as long as the student's questions get satisfactorily answered in the logical, semantical and epistemological sense indicated.

Another advantage to adopting the erotetic theory runs even deeper. Traditional research on teaching has relied on quantitative methods borrowed from the physical sciences. In the last decade or so the applicability of traditional methods to research on teaching and the broader domain of psychological and sociological research has been called into question. McKeachie (1974) and Cronbach (1975) have challenged the very possibility of universal laws of human behavior, at least any that endure across time. Others, such as Tom (1980, 1984), contend that the underlying metaphysical assumptions of natural science break down in the psychological and social domains. Still others, such as Fenstermacher (1979), doubt that the results of traditional quantitative research can be converted into a form applicable to the intentional classroom situation. The common denominator of these criticisms is that traditional methods are simply inadequate to deal with the goal-directedness and purposefulness of human social and linguistic interaction, and *a fortiori* less than adequate for the investigation and control of teaching.

In our everyday lives we often find it useful to conceive of our own actions and those of others as purposeful, goal-oriented activities that may only be understood by grasping the agent's intentions for (or *in*) undertaking an action or affirming a belief.

Everyday intentional actions (and interactions) commonly require a great deal of interpretation in order to be understood. One especially important instance of this is everyday conversation. Classroom dialogues are a natural extension of such conversations. And questions and answers lie at the core of these dialogues. One of the principle advantages of the erotetic theory is that it is capable of providing the logical structure that teaching dialogues obey in whatever setting they may occur. Logical structures do not decay; they are unaffected by the specifics of time or place. The enduring character of logical structure allows the formal properties of natural discourse to be determined in advance. Formal, logical predetermination does not translate into material *prediction* of either the content of teaching dialogues or when and where they may occur. Erotetic theory does allow the manipulation and *control* of such dialogues and even offers some suggestions as to how to initiate and direct question-answer discourse. The theory also makes it possible to *explain* why some forms of teaching discourse fail where others succeed. Erotetic theory is useful in the planning, implementation and evaluation of teaching. In fine, erotetic theory achieves most of the ends of traditional conceptualizations of teaching without carrying too much metaphysical baggage. Erotetic theory also avoids the pitfalls pointed out by Tom, Fenstermacher, McKeachie, Cronbach, Macmillan and Garrison and others. It does so by focusing its attention not on the "quantitative" aspects of teaching, but rather on the intentional.

Ericson and Ellett (1982) would replace traditional conceptualizations of research on teaching with something else - what they call "interpretive understanding," by which they mean "something akin to those categories of common sense by which we explain behavior in everyday life" (p. 494). We sympathize with their interpretive viewpoint, along with their additional claim that meaning only emerges intersubjectively. We agree too with their conclusion that the view of rationality prescribed by the new non-positivistic philosophy of science, and presumably any scientific theory constructed according to its tenets, is "semantic and pragmatic in [a] form that defies reduction to the admittedly elegant syntax of symbolic logic" (p. 503). But from this interpretivist conclusion it needn't follow, as so many seem to think, that "seeking understanding" is more important that "'having' knowledge, prediction and control" (p. 511); nor do we think that there is a need to abandon the quest for universal laws of human behavior. This is where erotetic theory and logic come in. Erotetic logic is a semantic and pragmatic (in the linguist's and logician's sense of the term) logic. Moreover, it is a rigorously intentional logic, one that provides interpretive understanding without making it necessary to abandon the quest for explanation, control and even prediction. The formulations of erotetic logic predetermine the structure of question-answer dialogues and show in part how they are to be meaningfully interpreted without specifying anything about their content.

Recall articles 4-9 of the Intentionalist Manifesto. At the center of our erotetic reconceptualization of teaching lies a reconstructed idea of causality. This idea differs in a number of ways from the traditional positivistic notion of causality borrowed largely from the Humean analysis of causation. One of our tasks will be to show how causality might be considered in intentionalistic contexts like teaching.

Let us conclude our introduction by quoting from another introduction, that of David Hume in *A Treatise of Human Nature*. There, Hume writes

> Tis evident, that all the sciences have a relation, greater or less, to human nature; and that however wide any of them may run from it, they return back by one passage or another. Even *Mathematics*, *Natural Philosophy*, and *Natural Religion*, are in some measure dependent on the science of man. . . . We ourselves are not only the beings, that reason,

> but also one of the objects, concerning which we reason. . . . We must therefore glean up our experiments in this science from a cautious observation of human life, and take them as they appear in the common course of the world, by men's behaviour in company, in affairs, and in their pleasures. Where experiments of this kind are judiciously collected and compared, we may hope to establish on them a science, which will not be inferior in certainty, and will be much superior in utility to any other of human comprehension.

Somewhere, somehow, this view of the primacy of the human sciences over the physical got turned around. It is not hard to fix the works of Mill and Comte as marking the turnaround. In what follows we seek to restore the original order at least a bit, by providing a rigorous theory of the intentional activities of teaching.

CHAPTER II

AN EROTETIC CONCEPT OF TEACHING

In this chapter we will try to provide as precise and rigorous an analysis of the concept of teaching as is possible without leaving the ordinary notion in the conceptual dust. We shall defend this analysis by pointing out some of its promise as the central feature of a full-blown theory of teaching, which we shall call an "erotetic" theory.

To teach someone something is to answer that person's questions about some subject matter. This is (roughly) the definition we shall try to clarify and defend. More fully and formally: any full-blown teaching claim[1] entails the further characterization of the teacher as answering a student's question about some subject matter.

Put thus baldly our contention is counter-intuitive. It does not match up with many contexts in which one might say that one person was teaching another. For example, such things as a teacher's praising and blaming students, assigning materials for study, and drilling students can all be subsumed under the characterization of the situation as teaching. Yet none of these (which we shall call "peripheral acts" of teaching) is prima facie analyzable as the answering of questions.

1. I.e., a claim that some person is teaching, has taught, etc., some other person some specifiable subject matter.

Teaching and Intellectual Acts

To see the point of our analysis, a short trip through the notion of "intellectual acts" is necessary. Recent philosophical analyses have stressed the importance to teaching of what Komisar (1968) called "intellectual acts," Green (1971) called "logical acts," and McClellan (1976) neologized as "lecting." These acts are exemplified by an open-ended list of verbs such as 'prove,' 'explain,' 'describe,' 'justify,' 'demonstrate,' and 'narrate.' Without rehashing all the arguments that have been put forth concerning the place of intellectual acts in teaching, we shall assert the conclusion: any characterization of an interaction between people as teaching requires that some intellectual-act characterization be lurking about. The test: try to imagine an act of teaching which did not implicate at least one of these acts in some way.

Bromberger (1965) provided the key to the type of analysis which we find most useful in explicating the notion of "intellectual acts." It hinges on the fact that a verb like 'explain' (which is the focus of Bromberger's attention) is an "accomplishment term", following the four-way classification of verbs found in Vendler (1957). This is important because of the temptation to consider 'teach' as (generically) a verb which exemplifies the "task-achievement" ambiguity.[2]

In a sentence like "Albert explained to Bertha why the tides rise and fall," 'explain' appears an "accomplishment term." It is worth

2. This has actually been an almost standard way of viewing the issue. See Ennis (1986) for a discussion directly relevant to our concerns, and Ericson and Ellett (1987) for further discussion of our approach. This type of analysis began with Gilbert Ryle (1949); philosophers of education latched onto it in the late 1950s. Notable among the latter are B. O. Smith (1961), and (later) Green (1971) and McClellan (1976). A critique of this type of analysis is presented in Macmillan and McClellan (1968). Needless to say, we find the task-achievement approach wanting. For a fuller discussion, see Macmillan and Garrison (1986), upon which this discussion depends heavily.

quoting Bromberger at some length on the distinction between activity terms and accomplishment terms:

> The difference between activity terms and accomplishment terms is readily seen when we compare their simple past tense. Both types have a simple past tense which implies that the continuous present was applicable at some moments in the past. The simple past tense of an activity term is applicable as soon as such moments have passed, and implies only the existence of such moments in the past. Aristotle walked. This implies that during some moments in the past Aristotle was walking. It does not tell whether or not Aristotle is through walking. The simple past tense of accomplishment terms implies more. It implies that relevant activities took place in the past, but furthermore that they have come to an end. And not to a mere stop, but to a conclusion. In other words, the simple past tense of accomplishment verbs entails that something has been finished, completed, that might, in principle, have been left unfinished, incomplete, that might have been stopped before coming to its logical end. . . .
>
> Thus accomplishment terms differ from activity terms in being associated with distinctions between what constitutes completion and what constitutes mere stopping, mere interruption. (Bromberger 1965, p. 75)

A full-blown teaching claim involves 'teach' as an accomplishment term. "Albert taught Bertha that Darwin was a gradualist" and "Albert taught Bertha how to swim" both entail that at some time in the past, Albert engaged Bertha in a set of activities which constituted his teaching her something. But this sets a new problem: what criteria characterize the completion of such interactions? A simplistic analysis might just say that it is Bertha's learning that determines the criteria of completion: i.e., completion of teaching is the student's getting what was taught. This answer has been given by Ennis (1986), among others. Presumably, any activity which brought about such a result would constitute Albert's teaching.

The element of truth in this answer is this: the criteria of completion and success in teaching are the same when teaching is seen as an accomplishment term. This is important to recognize, for it means that the teacher's responsibility may be broader than ordinarily thought. Consider the example of teaching someone how to swim: one might assume (as Ennis does) that success in teaching someone how to swim is manifested only in the learner's actually being able to swim. If something intervenes and the learner cannot swim, then the teaching has been unsuccessful. But when we search for reasons why that person cannot swim, we can see that some can be considered failures in completing the teaching, some in external factors. Bertha, we find, has not been attending to the demonstration of different moves which constitute swimming and cannot execute them as a result. This is a failure in teaching, since Albert was not teaching *her* how to swim; the teacher's responsibility extends to being aware of and taking care that the conditions of her getting the material are met. Clara, on the other hand, attends carefully, but gets a cramp and is paralyzed; she too cannot swim, but it is not a failure of teaching that makes it so. This diagnosis, it should be clear, depends only on the generic concept of teaching, and any analysis of teaching has to take it into account.[3]

But we want more than this, for there are various ways in which teaching can fail to be complete. It would be an erotetic failure if the wrong questions were answered by Albert, if his explanations and demonstrations to his students had little or nothing to do with staying afloat and moving through the water. And it would equally be a failure of the teaching if his answers did not relate to gaps

3. For a criticism of this point, see Ericson and Ellett (1987); a response is found in Macmillan and Garrison (1987).

in their knowledge of how to go about such things, i.e., if what he did did not speak to the questions that they ought to ask. The content of teaching can be wrong (or right) in both these ways. And these erotetic diagnoses can be made only with a concept something like ours: they enable us to see more clearly what is involved in teaching.

The erotetic analysis also helps us to see how other acts of teaching fit around its central logical core. Peripheral acts of teaching (e.g., assigning papers, directing drill, disciplining students, etc.) are usually accepted as teaching because of their contribution to the goal of the students' learning, when that is defined independently of the pedagogical interaction. But these acts can equally well be seen as contributing to the conditions favorable for engaging in the intellectual acts of explaining, describing, narrating, and so forth. The peripheral acts of teaching can always be seen as leading into or as supportive of the essential intellectual acts. The latter seem to be logically necessary, in the sense that we would deny that an interaction for which such a characterization was irrelevant was properly called teaching. On the other hand, the peripheral acts are not necessary in the same way. Socrates did not have to adjust the window blinds before explaining to Meno that his definitions of 'virtue' were inadequate, nor must a father assign homework when he teaches his child how to ride a bicycle.

On the other hand, we doubt that an elegant deduction from accepted premises could ever be deployed to "prove" the intellectual act view of teaching. But we doubt also that anyone could come up with a description of teaching as a set of actions that did not include an intellectual-act characterization. Even the rote drills that the teacher directs - the "exercising" necessary for the development of skills (McClellan 1976, pp. 128-130) - are related to some question of how to do something combined with the desire that the student become skillful at it. To describe what goes on in teaching without mentioning the intellectual acts is parallel to telling what happened at a bridge party without mentioning the bidding and play of cards, or to describing a round of golf without mentioning the clubs, ball or strokes made by the players.

Still, the notion of an intellectual act is vague enough that it cannot yet provide adequate guidance to the teacher or the necessary conditions for the pedagogical researcher. For one might turn to the standard philosophical literature on the logical forms of (for example) explanation and assert simply, "When one explains, one tries to approximate *this* form." Such a view would miss the interactive nature of teaching that our view emphasizes. It is not enough, either, merely to say that teaching is a type of interaction, for so are such disparate activities as playing catch and making love. Insofar as teaching is a type of interaction, characterizable as intellectual acts, what needs spelling out is the "logic" of the relationship - the rules and expectations that provide its form. These are not well captured by the ordinary metaphors and concepts found in the pedagogical literature.

Erotetic Analysis of Teaching

What distinguishes intellectual acts from other interactions among people? First, it should be noted that these are "speech acts" in the philosophical jargon (Austin 1962; Searle 1969). Intellectual acts, like many other speech acts, are "auditor-directed;" that is, these acts are to be taken as directed to an audience, to the students; not, "Newton explained the motion of the tides," but "Albert explained to Bertha why the tides rise and fall." What distinguishes the intellectual acts from such other auditor-directed verbs as 'warn,' 'promise,' 'order,' and 'request' is that the intellectual-act verbs require that the speaker have assumptions beyond the basic language-and-act understanding assumed of the others (Macmillan 1968, pp. 242-243). The speaker who describes, explains or narrates something to an audience must make assumptions about the audience's knowledge of and interest in the subject under consideration. One can warn without assuming that the hearer is interested in being warned, but one cannot explain the tidal flow or describe a process to someone without assuming that the hearer is in some particular intellectual or epistemological state with regard to the subject being discussed. The case of the promise to an infant demonstrates the distinction: If Martha promises her infant son that she will care for him forever, is it a full-blown promise? It is arguable - a defense can be given either way. But explaining the movement of the tides to that infant, knowing that he has no knowledge of the tides or anything related to them is absurd; no one could argue that Martha was explaining it to *him*, even if the same person would argue that she was "really" promising him the world. The cases differ in just this way - and it hinges upon the ability of the auditor to comprehend the explanation *as* an explanation.

So something about the auditor's - read "student's" - state of mind makes the difference and is at the heart of the logic of teaching. But how to capture this along with the interaction of student, teacher, and subject matter is the problem.

Here questions come into the picture - for the logic of questions provides a way of precisely defining the conditions for the different intellectual acts. Each intellectual act is directed to a corresponding question: for explaining, "Why?" or "How?"; for narrating, "What happened?"; for describing, "What does it look like?" and so forth. For moral and practical reasons, one might hold, as does McClellan (1976, pp. 111-113), that teachers only teach when the form and content of the interaction consists of the student's explicit questions being answered by the teacher or through some cooperative effort of teacher and student. But this won't do if we are not arbitrarily to rule out many examples that would be accepted by any observer as teaching.[4] If we are cor-

4. One criterion for an adequate conception of teaching is that it must fit ordinary contexts in which 'teaching' is the verb of choice for an accurate description; and it is not a necessary condition of such descriptions that they show explicit questions being asked and answered.

rect in holding that teaching is constituted of intellectual acts, an explicit-question interpretation of these acts must be rejected; one may describe a scene to someone without being asked to do so, or explain something to someone who isn't even aware of wanting to know why it happened, or who would not ask the question, "Why did it happen?"

Rather than asserting that the question must be explicitly asked, or assuming that it would be if the student just thought about it, a different approach is necessary to clarify the logic of the interaction. It is not that the student does ask the question, or that the teacher even believes that the student would ask the question, but rather that the teacher believes that in some sense, the student ought to ask the question. The person who describes a scene or a painting must believe or assume that the auditor's present state of knowledge and perceptual stance makes the asking of the question at least relevant and perhaps necessary. It would be senseless for someone to go through the motions of describing a painting to an auditor who could describe the painting as clearly and cleverly as the describer and the latter knows this.

This indicates the logical importance of questions to teaching; the teachers' lesson, discussion, or lecture is the answer to a question or set of questions that the students are intellectually in a position to ask. Strategically (but not necessarily) the teacher might want to make the question explicit, to show the students why they ought to ask that question given what they already know, but strategy is not the same as conception - even if it presupposes a conception like this. A teacher with a different style, faced with different sorts of students, might leave the explicit statement of the question until the answer has been given - perhaps to show the relevance of a particular point in a lesson by emphasizing that it only makes sense in the context of a particular background of knowledge and belief, that it is related logically to a particular question.

Questions can be categorized according to the type of intellectual predicament their answers can be used to overcome.

We have put it too simply, of course, for not all intellectual predicaments get expressed as questions. One can be in such extreme ignorance of a phenomenon that he or she could not even ask questions about it. "What is Newton's third law?" is a question beyond the possibility of being raised by the average five-year-old, yet we might want to teach such a child what the law is, or to explain its meaning to the child. Explaining the law might require a great deal of explication of unfamiliar material, but it could be done.[5]

The notion of "intellectual predicaments," then, is wider than that of questions or even of puzzlement. It includes those gaps in knowledge, those ignorances, that a questioner could not even raise, as well as those that lie behind questions explicitly asked. The theory of intellectual acts correspondingly must go beyond the answering of questions into the great variety of intellectual predicaments.

Bromberger (1965) analyzed 'explain' from a point of view similar to this one. Let us summarize the two predicaments which he cites as those which a "tutor" might (disjunctively) try to overcome in explaining something to a "tutee"; the tutor must assume that the tutee is in either (or both) of two predicaments with regard to a why-question about the thing being explained: (1) the tutee views the question as sound (i.e., as admitting of a right answer), but cannot think of, imagine, or excogitate a possible answer to which, in his present state of knowledge, he has no decisive objections. (This Bromberger calls a "p-predicament.") (2) The question is sound, and the answer to it is beyond what the tutee can conceive of, imagine, or remember. (This is called a "b-

5. Bruner (1960) put forth the "hypothesis" that "Any subject can be taught effectively in some intellectually honest form to any child at any stage of development" (p. 33). The erotetic approach may give new meat to the slogan's bones.

predicament.") These two predicaments differ. A person in a p-predicament must ask the question (or be able to ask it, at any rate) and must know enough to reject some answers to the question, but the question may be unsound, or his objections may be based of false premises. The person in a b-predicament might not even be able to ask the question, but the question is sound.

The total list of conditions in Bromberger's final "hypothesis" includes more than the predicaments; it includes also the explainer's knowledge, and the actions performed by the tutor in order to relieve the predicament - actions that may include showing the predicament to be unjustified, or the question to be unsound.

One must ask whether this way of analyzing the act of explaining something to someone else has some justification beyond its neatness and beyond Bromberger's justification, which is that it helps to make clear both what "an explanation" of something is and also how a theory can be said to explain something. Bromberger's view has two major advantages for pedagogical theory. First, it does not limit explanations to a single logical pattern, as does the traditional covering-law model. Second, and closely related, it shows how material which is not essential to deductive patterns comes in - one sometimes must explicate a whole theory in order to explain a phenomenon to a given student. Bromberger's way of analyzing intellectual acts does justice to the ordinary language distinctions that are made by using these terms. One can attempt to analyze them purely from the point of view of form or of content, of course. But such attempts fail finally by narrowing the concepts, for one can describe in odd or extraordinary ways, evaluate by using metaphors, and so forth. And a purely formal analysis might miss the unusual or odd.

In sum, as intellectual-actor, the teacher must assume that in regard to the subject to be explained, described, or demonstrated, the students ought to ask a question; we will call this an "epistemological ought," and the intellectual state of the students an "intellectual predicament" with regard to the subject matter. The epistemological ought is defined by the intellectual predicament.

It is important to recognize that the "ought" involved in this view is not to be thought of as a moral duty or a moral ought, for it has to do with the epistemology of the situation: the students being taught have to be viewed as being in a particular intellectual predicament with regard to the subject in question and hence cannot be thought of as being in a position where they ought to ask that question without having some knowledge or beliefs about the thing questioned. This is best seen with simple examples. The person who asks, "Who lives in that building?" implies that he or she believes at least the following things:

1. That it is a building he/she sees and not a stage set, for example, or a mirage.
2. That people live in buildings.
3. That someone lives in that building.
4. That the person questioned knows the answer to the question.
5. That it matters to him or her who lives in that building. (This point was suggested to us by Professor James E. McClellan, who also thinks that "matters" should be defined materially.)

This suggests that a formal concept of teaching can be developed which includes all three aspects of the three-part interaction: teachers, students and subject matter. For in this analysis, the teacher's actions are logically relevant to the students' intellectual states with regard to a particular subject matter.[6] And the

6. A person who gratuitously says to his guest, "Mabel Smith lives in that building," can be viewed as answering a question that the guest might ask about it. Note the problem here - any comment, description, statement,

logic that ties them together is erotetic logic - the logic of questions.

Erotetic logic attempts to deal with at least the following two problems: (1) What are the presuppositions of different types of questions (presuppositions here being the types of implication noted above); and (2) What makes a given response a proper answer to a particular question? (See Hintikka 1976, for the fullest discussion.) For a beginning, we may be guided by the old slogan, "The question determines the form of its own answer." Although imprecise, the slogan points to the fact that the relationships between questions and their answers are logical or semantical.

This in turn suggests the importance of the erotetic analysis to pedagogical theory: *It is the intention of teaching acts to answer the questions that the auditor (student) epistemologically ought to ask, given his or her intellectual predicaments with regard to the subject matter.* Insofar as these questions can be put in a clear and unconfused way, the questions will have exact and determinate semantical (and possibly syntactical) content. The correct answer to a properly posed question must be couched in the terms within which it was asked. We may say with some confidence, then, that intentional intellectual acts of teaching hit their marks when they satisfy the semantical and syntactical demands of the questions the students epistemologically ought to have asked given their intellectual predicaments. Insofar as this is done, it is assured that the teaching hits its mark with such accuracy, rigor and precision

etc., can be viewed as answering a question, if this is right. But if everything is covered, isn't the whole theory vacuous? Perhaps it is not necessary to try to defeat this extension of the notion of questions-and -answers dialogue, however. Warnings, promises, and so forth do not presuppose possible questions; the erotetic view gives a criterion of relevance, and hence an explanation for why people don't understand our gratuitious statement: they don't see the point of them. Seeing the point isn't too much more than seeing what possible questions the statement is relevant to. Bores, exuders and many teachers miss this point. See Komisar and Associates (1976).

that the issue may almost be turned over to the erotetic logician for further study and without further empirical ado.

But questions still remain unanswered. It is still not settled that all instances of teaching involve question-answering (Ennis 1986). It should be recognized that we have to defend not merely the erotetic analysis of the concept of teaching, but also the intellectual-act assumptions spelled out above, for our analysis gets its foothold through that move; it is a way of spelling out what is involved in explaining, narrating, proving and showing (. . . etc.) something to someone else. One apparent counterexample to the foregoing erotetic analysis of teaching arises out of the underlying intellectual act assumptions: *teaching how*. This is important to deal with, since question-answering seems most relevant where the teaching is "propositional", i.e., where the subject of concern is statable as a (declarative) sentence.[7] Our contention - to be defended - is that the erotetic analysis fits other cases just as well. We shall deal with some counterexamples suggested by Ennis (1986).

The first question: When is it appropriate to teach someone how to swim?[8] Reverse the question to see its complexity: When is it inappropriate to teach someone how to swim? First, of course, when the student already knows how to swim. This is a feature of the generic concept of teaching: one cannot teach a person something he or she already knows. Second, when the student's

7. The breakdown assumed here distinguishes between "propositional" teaching and "teaching how", "teaching [someone] to . . . ", as well as "teaching the . . . ", "teaching when . . . ", and various other "wh" constructions. See Scheffler, 1965, and Komisar, 1967, for different approaches to the same breakdown.

8. Swimming seems to be a favorite example of philosophers of education; this may be because swimming involves such a myriad of activities, long practice, hands-on in-water demonstration, and so forth. Howard (1982) uses another example: teaching students how to sing. That too is complex and difficult.

experience is such that he or she could not ask even the general question, "how do I propel myself through the water?" much less such subquestions as, "Must I keep my mouth out of the water at all times?" A Saharan, perhaps, would have no notion of what swimming was all about, having never seen a body of water large enough to immerse himself in. Between these two cases are a large range of possible questions to ask about swimming. The student must have some questions about the procedure or we cannot begin to teach.

Consider a parallel case in *teaching that*: when Bertha knows or believes already that Darwin is a gradualist (Ennis's example), Albert cannot teach *her* that fact. But he couldn't teach another student, either - the one who is totally ignorant of the matter: "*Who* is a *what*?" asks this student uncomprehendingly. In order for Albert to teach her, he would have to make sure that the proposition could make sense to her, that the question, "Was Darwin a gradualist?" arises because of her present state of knowledge.

Normal swimming students, though, have some questions. In order to answer the general "How does one (do I) swim?" we break the activity down into subactivities, each of which involves the same logical relation between question and answer: "How can I breathe in the water?" "How does one move her arms and legs?" and so forth.[9] Our answers may take the form of a demonstration and the direction of practice to make sure that the students get some feel of what it is like to make such moves.[10]

9. Whether or how a complex activity like swimming (or singing) should be broken down into subquestions is a problem for empirical and logical investigation, not something given by the erotetic analysis of the concept of teaching. The example is used for argument, not as a recommendation to teachers. See Chapter VI, below, on strategies of teaching.

10. Again, Howard (1982) shows us something here: a lot of what goes on in teaching singers is intended to get them to know what it *feels like* to produce a head tone or breathe from the diaphragm (pp. 168-173).

Suppose that the questions have been answered, that they are true both to the activity of swimming and to the students' intellectual (and physical?) predicaments with regard to swimming. What else is left to the *teaching*? Directed practice, perhaps. But directed practice itself may best be seen as the attempt to answer questions about how to do things: "What happens if I kick this way rather than that?" "How does one do the Australian crawl?" and so forth.

What else is left? "The student's learning," we imagine the critic saying, with a cry of frustration. "Her actual swimming or being able to swim! We shouldn't claim to have taught someone how to swim unless she *can* swim!"

But what does that amount to? There's a problem here, but it is a problem with the notion of a learned ability, not with the concept of teaching. When should we say that someone has been taught how to swim? Much will depend upon the context of the claim; if the situation is desperate enough, we will accept very little by way of expertise. The flood waters are rising; the Saharan does not know how to swim; "Do this, and that, and the other thing," we say, demonstrating some basic swimming moves, getting him to try them out, perhaps. No time here for the fine points of the butterfly stroke. Have we taught him how to swim? If we have answered the Saharan's questions adequately, yes. Suppose he sinks? Why did he sink? Was it a failure of teaching, or did he panic? Perhaps it *was* a failure of teaching: "He didn't understand that one swims *in* the water rather than *on top* of it, so did the wrong things." (See McClellan 1976, p. 130.) His drowning is not *necessarily* a falsification of our claim "We taught him how to swim."

In fact, not much seems left to teaching when the questions have been answered. The development of the habits of (good) swimming may take more time than the episodes of teaching, and the supervision of the practice that develops those habits can be seen as an appropriate task for the teacher of swimming. Our contention would be that those practicing episodes should be seen as question-answering just as would be the "intellectual" episodes of *teaching that* or *teaching the* .... This is not verbal magic - merely a closer analysis of what actually goes on in teaching.[11]

One word on *teaching to.* Consider this anecdote: "My father taught me to dim the lights before an approaching car reaches the top of an intervening hill. He only said it once, but I've done it ever since." This seems to be a clear case of *teaching to*, yet it involves only one episode, one question answered. The teller of this tale ought to have asked (but perhaps did not) "When ought I to dim the lights?" and the answer was clearly given. Teaching someone to do something almost always involves ways of acting in particular circumstances, doing the right thing given the context. Developing the habit may involve long practice, considerable demonstration of how to do it, and so forth. But the "disposition" of dimming the lights can be seen clearly as the answer to a particular question. General dispositions, like "being honest," may be more difficult, but they too can be seen to be answers to questions: "How ought I to act, in general?" Answer: "Honestly." The rest is detail, practice, courage. Someone who has not learned to be honest may indeed have been taught to be honest; the teaching claim has to do with the interaction of the teacher and the student with respect to the type of conduct under consideration, and only indirectly with the achievement of long-term dispositions.

If these examples ring true, there should be little difficulty in seeing the power of the erotetic analysis of teaching or its ability to handle difficult cases. Our concern is with the nature of the relationship between teachers, students, and subject matter; that relationship has the logical structure of question answering. At-

11. The exasperation of teachers when students don't seem to follow their directions or explanations becomes clearer. They have *taught* the students what to do, but the students seem not to do it. Whose fault is that? It will differ from case to case.

tending to the logical structure goes a long way toward answering questions about what teaching is all about. It does not, as Pendlebury (1986) so aptly points out, tell us how to teach or exactly what moves to make in what order. There is a great deal of room for empirical investigation here, even a place for process-product research, perhaps.[12]

The "Epistemological Ought"

But there are even more unanswered questions. Some additional features of the erotetic analysis of teaching must be sketched in before the full power of the theory can be considered.

At the center of this analysis is the notion of the epistemological ought, the question that a person, given a particular intellectual predicament, ought to ask concerning the subject matter. Several points need to be made concerning the nature of the epistemological ought, since it is open to various interpretations and misunderstandings.

First, as already stated, the epistemological ought is not *explicitly* a moral ought. Pendlebury (1986), however, wonders whether in educational contexts epistemological oughts aren't always grounded in moral oughts. In one sense at least, they are, but it is an ultimate rather than a proximate grounding. Math students, for example (Pendlebury's), ought to ask, "Does 'difference between' mean the same as 'is less than'?" only in the context of learning mathematical languages and procedures. The context in which such a question comes up is one in which we have decided that they should be taught such a language. The decision to teach such things to children is very serious morally; we are deciding that they should become people who know mathematical language, not people who are ignorant of it. Whenever we make such decisions, especially in school contexts, we are putting a stamp of approval on the material being taught. Within the global context, however,

12. But see Chapter III, below, for further consideration of this issue.

the moral dimension is or can be taken for granted; once we have entered upon the task of teaching mathematical languages, the subquestions become more epistemological than moral.[13]

Other questions come up, of course; there are an indefinite if not infinite number of questions that students ought to ask concerning a particular subject matter. The teacher must select among these possibilities as the lesson develops. But the problem is not with the erotetic analysis of teaching, but with the complexity inherent in the relations among the student, teacher, and subject matter. In fact, the issue seems clearer when considered erotetically than when it is seen merely as a general problem for teachers' planning. Here's the point: when teachers develop plans for their students, they must consider not only the students' current state of knowledge and the structure of the subject matter, but also where any particular subject or part of a subject will take them. The answers to some questions will just shut off discussion. "Is abortion wrong?" "Yes." This stops the discussion. "What is wrong or right about abortion?" extends it, because new questions arise, new topics for discussion, new dimensions of the justification or condemnation of abortion. To know which questions to answer requires that the teacher have considerable knowledge of the students and of the subject matter; without this, the teacher cannot know where a particular question or answer is going to lead.

A full discussion of this would require attention to the structure of the lesson that would go beyond the simple answer to a particular question. Answering one question, for example, will es-

13. These decisions about curriculum and the ultimate aims of education are in practice political as well as pedagogical; that this is so does not weaken the erotetic theory, but rather points to the importance of being absolutely clear about all of the factors which enter into pedagogical decision-making. The erotetic analysis emphasizes only one dimension; a more politically oriented approach might attend to the ways in which this theory could be misused or abused in the practice of schools. But that approach should have the clarity that erotetic analyses provide for the diagnosis of evil.

tablish a presupposition for other questions, which then might be the topic of further concern. These are decisions that teachers *must* make; students' background knowledge and the structure of the subject matter may greatly constrain what questions ought to be asked, but alone they seldom determine a unique question. The teacher's creative autonomy is not only desirable but necessary in erotetic teaching. Erotetic logic helps in seeing the structure, but, as we will see below, it does not answer all the questions involved in determining what to teach.

Second, and very importantly, it must be re-emphasized that any question-asking-and-answering interaction is logically related to the asker's state of knowledge regarding the subject. It has long been recognized that a question makes sense only if in some way the questioner is puzzled by the thing he or she asks about (Dewey 1933; Hospers 1946). The significant thing about puzzlement is not some psychological "itch" which might accompany it, but rather the relationship between the person's understanding and the object which puzzles. Showing that Bertha is puzzled does not require proving that she has some peculiar feeling concerning the object of puzzlement. We could imagine her being puzzled by stripes on a wall and by stripes on a tiger without having to imagine that her feelings were the same in both cases. What we could not imagine would be that she was puzzled about the stripes in either case but had no unanswered questions about them. Puzzlement is defined by the types of questions that a person can and cannot (*logically* cannot) ask about the object of puzzlement. (See Macmillan, 1968, for a fuller discussion.)

Closely related to this is a third point about the epistemological ought: it does not totally determine how a teacher should proceed. Consider this feature of the analysis: for any body of subject matter (presumably analyzable into concepts and assertions), any given student will be in one of an indefinite number of possible intellectual predicaments. These range along at least two dimensions - first, from total ignorance to maximum knowledge of the subject, and second, with regard to any particular part of the subject, an indefinite number of possible intellectual predicaments: the teacher selects from among them, and this selection will be based not solely upon the intellectual predicament; the teacher will select that question to answer which has other possibilities - e.g., the one that is broadest in its implications, or the one whose answer will provide the most information. The epistemological ought is related to other matters than the student's current knowledge, for it is the relation of the student, teacher and subject matter that is at stake; and the decision about what possible questions will further promote and expand the relationship is one that the teacher presumably makes from a position of greater knowledge than the student.

Again the logic of questions is relevant; erotetic logicians are moving into the discussion of such issues as the intellectual presuppositions of questions (Bromberger 1965; Martin 1970), the conditions governing complete and partial answers (Hintikka 1982), and the use of questions in information-seeking situations (Hintikka and Hintikka 1982). The power of different kinds of questions to motivate further investigation and learning has also attracted attention. Developments in this rather esoteric field should throw light on the more mundane problems that teachers face every day.

CHAPTER III

A PHILOSOPHICAL CRITIQUE OF PROCESS-PRODUCT RESEARCH ON TEACHING

Our goal in this book is to develop an approach to teaching that is both empirically rigorous and intentional. As indicated in Chapter I, the dominant tradition of research on teaching has tried to achieve empirical rigor by avoiding the intentional components of teaching. In the present chapter, we will employ some principles of postpositivistic philosophy of science (Garrison 1986) to show how and why the various stratagems employed within the dominant research tradition to finesse the problem of intentionality have failed and why it is time to begin exploring new alternatives such as those suggested by the erotetic concept of teaching.

In the preface to the *Handbook of Research on Teaching*, Nathaniel Gage (1963b) noted that research on teaching had "lost touch with the behavioral sciences," had, in effect, become divorced from the rigorous approach that scientific endeavors entail. That handbook was intended to provide a foundation for and review of research on teaching; one of its lead articles was Gage's own "Paradigms for Research on Teaching," (1963a) which set forth the basic structure of the paradigm that he saw as necessary for theoretical and practical development in teaching. He was not explicitly developing a theory of teaching (if such there be), but rather the "paradigm" that would provide the framework for developing such theories: "Paradigms are models, patterns, or schemata. Paradigms are not theories; they are rather ways of thinking or patterns for research that, when carried out, can lead to the development of theory" (Gage 1963a, p. 95).[1] A remarkable quotation this, for it was written before Kuhn (1962, 1970) introduced the term 'paradigm' in this sense into the literature and long before the introduction of further subtleties by such writers as Lakatos (1970) and Laudan (1977). Gage understands well that one should be able to distinguish between the methodological and ontological commitments of what he calls "paradigms" and the findings of particular research studies. Indeed, he set out to develop the general metaphysical and methodological assumptions of his research tradition, which he calls there "the criteria of effectiveness paradigm" for research on teaching but which has come to be known as the "process-product" paradigm.

In this chapter we will examine the progress of the process-product paradigm over the last twenty years, working within the same tradition of philosophy of science that Gage suggested. Our examination is motivated by a suspicion that all is not well within this tradition. The difficulty is to see whether our suspicions are well grounded and to see what follows if they are.

1. Gage continues: "Paradigms derive their usefulness from their generality. By definition, they apply to all specific instances of a whole class of events or processes. When one has chosen a paradigm for his research, he has made crucial decisions concerning the kinds of variables and relationships between variables that he will investigate. Paradigms for research imply a kind of commitment, however preliminary or tentative, to a research program. The investigator, having chosen his paradigm, may 'bite off' only a part of it for any given research project, but the paradigm of his research remains in the background, providing the framework, or sense of the whole in which his project is embedded" (1963b, p.95).

For reasons that will become clear as we proceed, we do not believe that the process-product paradigm can be rejected for one reason alone; we do believe, however, that a combination of reasons weighs heavily in favor of searching for new approaches.

Gage Presented

Gage (1963a) thought that theories of teaching had been relatively neglected in past educational research. In diagnosing this neglect, he found two possible reasons: first, that learning as viewed by psychologists is a much more general phenomenon than teaching, hence worthy of more time and effort; second, and more important, is the relation between theories of learning and teaching.

> Another reason for the relative neglect of theories of teaching may be that they have been seen as unnecessary on strictly logical grounds. This position - itself perhaps a kind of theory of teaching - amounts to saying that, if we have an adequate theory of learning, then the teacher must of necessity act upon that theory, without employing any separate theory of teaching. The teacher, if he is to engender learning, must of necessity do what the theory of learning stipulates as necessary for learning to occur. Teaching must thus be a kind of "mirror image" of learning. This conception of what a theory of learning implies for teaching may explain the neglect of theory of teaching by psychologists. (1963a, p. 133)

In place of this view concerning the locus of learning theories in research on teaching, Gage proposed that such research be guided by and aimed at a theory of teaching parallel to a theory of farming:

> Farmers need to know something about how plants grow, and how they depend on soil, water, and sunlight. So teachers

> need to know how children learn, and how they depend on motivation, readiness, and reinforcement. But farmers also need to know how to farm - how to till the soil, put in the seed, get rid of weeds and insects, harvest the crop and get it to market. If our analogy applies even loosely, teachers similarly need to know how to teach - how to motivate pupils, assess their readiness, act on the assessment, present the subject, maintain discipline, and shape a cognitive structure. Too much of educational psychology makes the teacher infer what he needs to do from what he is told about learners and learning. Theories of teaching would make explicit how teachers behave, why they behave as they do, and with what effects. Hence, theories of teaching need to develop alongside, on a more equal basis with, rather than by inference from, theories of learning. (1963a, p. 133)

The new idea here is that theories of teaching must be relatively autonomous vis-à-vis other pedagogically relevant theories. Learning theory is to take its rightful place in the discussion of teaching, but that place is not to be the overarching theory from which teachers and educational researchers deduce methods of teaching or conclusions about effective teaching. Gage is not entirely clear what the place of learning theory should be, but at any rate it should have a subordinate role in research aimed at determining the characteristics and activities of the effective teacher.

For the purpose of the paradigm and its constituent theories of teaching is "to discover what makes a good teacher" (1963a, p. 114). From this task grows the basic frame of research.

> As soon as the idea of effectiveness enters the research, the question of a criterion of effectiveness is raised. The paradigm has then taken the following form: Identify or select a criterion (or set of criteria) of teacher effectiveness. This criterion then becomes the dependent variable. The research

> task is then (1) to measure this criterion, (2) to measure potential correlates of this criterion, and (3) to determine the actual correlations between the criterion and its potential correlates. In short, variables in research on teaching conducted according to the "criterion-of-effectiveness" paradigm have typically been placed in two categories: criterion variables and potential correlates. (1963a, p. 114)

The "criterion variables," Gage notes, may range across a wide array of possibilities, from the teacher's effect on pupils' achievement and success in life through a superintendent's satisfaction with the teacher (1963a, p. 117). But he proposes attention to "micro-criteria" of effectiveness, since there are so many possible conceptual and practical difficulties with more general criteria. "Rather than seek criteria for the overall effectiveness of teachers in the many, varied facets of their roles, we may have better success with criteria of effectiveness in small specifically defined aspects of the role. Many scientific problems have eventually been solved by being analyzed into smaller problems, whose variables were less complex" (1963a, p. 120). Gage finds the basic methodological structure of his paradigm in personnel-selection research. "Whether the purpose has been to select college students or clerical workers, clinical psychologists or airplane pilots, the same paradigm has prevailed: get a criterion and then find its predictors" (1963a, p. 115). The "ultimate" criteria that he seems willing to consider are always changes in pupils, either while in school, upon completion of school, or even in later years. The product in process-product research is change in students.

The plausibility and attractiveness of the process-product paradigm depend upon its simplicity: the tasks of the researcher are clear, consistent with existing research traditions in education and related fields of social science, and directly related to the improvement of teaching. The basic methodological procedures of variable analyses fit neatly into well-established statistical traditions as well.

At the metaphysical center of Gage's paradigm lie several assumptions essential to any science. Among these are (1) an assumption that at the very least, nature is uniform, i.e., that the objects of investigation are consistent in their existence and behavior. This assumption of the uniformity of nature is essential for any science that seeks to develop nomological laws. (2) The principle of causality is also assumed to hold - at least as far as believing that it is possible to give causal explanations of the phenomena under investigation, at most to believing that all events are caused. (3) A belief that our knowledge of the natural world must ultimately (and perhaps proximately) depend upon our experience of the world, as apart from "rationally" determined laws. (4) An assumption that the number of circumstances that pertain to the causation of particular natural phenomena is limited and knowable. Finally, (5) that only phenomena that admit of quantification (or better still, measurement) are fit for scientific inquiry.

It can almost be said that Gage's methodology determines his ontology: the method that he sees as central to the tasks of research on teaching - i.e., the statistical analysis of the relations between quantitatively stated variables - entails that whatever variables are studied must be stated as measurements of some sort. This is, of course, an extension of the long-standing tradition of educational research summarized by E. L. Thorndike in 1918.

> Whatever exists at all exists in some amount. To know it thoroughly involves knowing its quantity as well as its quality. Education is concerned with changes in human beings; a change is a difference between two conditions; each of these conditions is known to us only by the product produced by it - things made, words spoken, acts performed, and the like. To measure any of these products means to define its amount in some way so that competent persons will

> know how large it is, better than they would without measurement. To measure a product well means so to define its amount that competent persons will know how large it is, with some precision, and that this knowledge may be conveniently recorded and used. This is the general *Credo* of those who, in the last decade have been busy trying to extend and improve measurements of educational products. (1918)

Note, however, that where Thorndike was concerned to measure educational products and was working out the methodological necessities for that, Gage assumes that that problem is taken care of and turns his attention to the measurement of the processes that are the focus of his research tradition. While earlier researchers had been concerned with educational products, the full development of the other side of the equation was now to be the central task.

Throughout the history of process-product research on teaching, it is clear that what is being sought are the causes involved in effective teaching - i.e., the things that a teacher might do in order to cause the criteria of effectiveness (the product) to come to be, whether those be changes in students, new beliefs, or whatever. Gage is actually less explicit in 1963 than in later works. By 1966, he criticizes the work of Smith and Bellack as not clearly showing how their studies are (causally) relevant to teaching - and this won't do. "Not everything that teachers do is relevant to the purposes for which we study teaching. We typically do not concern ourselves with how teachers scratch their heads, hold their chalk, or cross their legs, for the simple reason that we assume such behaviors to be irrelevant to the kinds of learning the teachers bring about in pupils" (1966, p. 35).That this is indeed a causal comment is emphasized by Ennis:

> Deliberate education is the attempt to bring about certain changes in students. The answer to the question 'How can

> we bring about certain changes in students?' either is or assumes a causal generalization, because bringing some thing about is causing it to come to be. It is important for educators to have at least probable knowledge about what will be caused by certain factors and what will cause certain factors (Ennis, 1982, p. 25).

More explicitly, in 1978 Gage differentiates between mere "predictive" relationships of variables and "causal" relationships. In the process-product approach to research on teaching, he says, "We search for 'processes' (teacher behaviors and characteristics, in the form of teaching styles, methods, models, or strategies) that predict and preferably cause 'products' (that is, educational outcomes in the form of student achievement and attitude)" (1978, p. 69).

This brief presentation of Gage's "paradigm" for research on teaching is greatly oversimplified. We have tried to emphasize the aspects of the paradigm that form the basic assumptions of what Laudan calls a "research tradition" or the "largest identifiable unit of science." "A research tradition is a set of general assumptions about the entities and processes in a domain of study and about the appropriate methods to be used for investigating the problems and constructing the theories in that domain" (Laudan, 1977, p. 81). Our discussion is so general because we want to consider the effects of these assumptions upon the progress of research in this domain over the last twenty years. It is to that task that we now turn.

Gage Engaged

Although it is easy to stand back from a research tradition like Gage's and take potshots at individual assumptions and weaknesses, such potshots could never be conclusive in the decision to accept or reject it. Any such tradition will eventually build up a body of anomalies that seem intractable. From the outside, these may seem to call for outright rejection of the whole shebang, while from the inside they may be perceived merely as interesting problems to be solved by further research or ad-hoc theoretical tinkering. As long ago as 1906, Pierre Duhem emphasized that scientific theories not only have complex internal structures but also interact with one another in such logically complex and interpenetrating ways that it is impossible to isolate any single theory in order to perform a crucial experiment to verify or refute it in any incontestable way (Duhem, 1906). It is always possible to save an apparently refuted theory by the judicious addition of suitable ad-hoc hypotheses that either bear the weight of the anomalous instances or, in conjunction with the rest of the to predict the otherwise anomalous results. We cannot, therefore, expect a knockout blow to drive a theory (or research tradition) from the scientific ring. But we might expect, with Duhem, that awkward or inelegant theories or traditions might fade from the scene as a result of their own inadequacies.

This point has been made in many ways in recent philosophy of science. (See, for example, Kuhn 1970; Lakatos 1970; and Laudan 1977.) Tracing the historical development and demise of a scientific research tradition will show how various kinds of problems led eventually to the development of new and different assumptions about the best methods of research, about the basic objects to be studied, and about the point of the study. Looking back, then, one can see why certain traditions failed.

But the question at hand is whether we can stand in the midst of things, as it were, to see how a particular tradition is doing. Can the insights of the "new philosophy of science"[2] be brought to bear on current, more-or-less living traditions? Perhaps by considering current anomalies, criticisms, and defenses of a tradition, its

2. The term comes from Brown (1977).

viability can be assessed. Such an assessment might show ways in which the tradition can be improved, or it might suggest that there are other directions that might be more fruitful.

It is this task that we shall take on in the rest of this chapter. There are problems with the process-product tradition of research on teaching; but there are also successes to be accounted for, and some interesting defenses of the tradition. Gage stands us in good stead, again, for in his book *The Scientific Basis of the Art of Teaching* (1978) Gage attempts to assess his own tradition from within, answering criticisms and proposing new ways of interpreting its findings. He ends on an optimistic note, suggesting that the "normal science" of research on teaching is process-product research and that the future could be bright with only more support for and work within the framework of that tradition.

Empirical Anomalies

The major problem facing process-product research on teaching is, at least in its own eyes, empirical productivity: research conducted under the paradigm has yielded results that are at best inconsistent. Gage noted the problem in 1963: "Research by this paradigm has been abundant; hundreds of studies, yielding thousands of correlation coefficients, have been made. In the large, these studies have yielded disappointing results: correlations that are non-significant, inconsistent from one study to the next, and usually lacking in psychological and educational meaning" (1963a, p. 118). But the same problem was evident in 1978; Gage begins his discussion of the positive side of matters by citing negative reviews: "I am now entering into matters on which most writers' conclusions over the years have been negative; mine and some others' are positive. Most reviewers have concluded their reports by saying that past work has been essentially fruitless. Such discouraging characterizations of previous findings go back at least 25 years and are still being repeated in present-day publications" (1978, p. 24). The criticism had been put trenchantly by

Doyle: "Reviewers have concluded, with remarkable regularity, that few consistent relationships between teacher variables and effectiveness criteria can be established" (Doyle 1978, p. 165).

There are two semitechnical points about these criticisms. First, most studies done within the process-product tradition have shown relatively low correlations between teacher behavior and student achievement - in the range of $\pm.1$ to $\pm.4$; second, perhaps because of small sample sizes, the significance of the findings has rarely reached that expected within statistically adequate studies (Gage 1978, p. 26). In general, the findings have not been adequate in themselves to support strong conclusions about correlations between scientific processes and products. In a very serious sense the process-product tradition has not been productive. Doyle suggests also that the "productivity issues" are associated with unsolved methodological problems "that have impeded attempts to compare studies, integrate findings, or apply results to teacher education" (1978, p. 165). He suggests further that "two-factor" causal analyses - i.e., where teacher behavior and student achievement are the only two factors considered - overlook the importance of the "mediating processes" of student activities and of the "classroom ecology" that he tries to develop as a competing paradigm for process-product research (1978, pp. 170-188).

Beyond methodological problems, Doyle notes that the productivity issue is exacerbated by the fact that there are "few theoretical grounds for selecting variables for interpreting available findings" (1978, p. 165). It appears that the very generality and simplicity that make the process-product paradigm so accommodating also lead to serious gaps in its effectiveness as a guide for generating adequate theories or even categories for investigation.

Gage recognizes that the productivity issue is the single most important problem facing the process-product tradition. His response to criticisms like Doyle's draws heavily upon Glass's work on meta-analysis of data (Glass 1976, pp. 3-8; 1978, pp. 351-

79; Glass, McGaw and Smith 1981). Conceding that individual studies typically display rather small correlations, Gage argues that by "clustering" studies together and applying meta-analytical techniques it is possible to achieve significant results at a different (meta) level.

> Thus the path to increasing certainty becomes not the single excellent study, which is nonetheless weak in one or more respects, but the convergence of findings from many studies, which are also weak but in many different ways. The dissimilar, or unreplicated, weaknesses leave the replicated findings more secure. Where the studies do not overlap in their flaws but do overlap in their implications, the research synthesizer can begin to build confidence in those implications. (1978, p. 35)

With this Gage brings the matter to rest.

A full philosophical treatment of the significance of meta-analysis of data will have to wait for another time, but some remarks can be made here. Recomputing and combining results to make them seem more significant is economically a good move: one can see trends if not firm conclusions about the degree of relatedness of the factors studied. Furthermore, this type of analysis can dampen the effect of higher-level interactions, thereby suggesting the causal structures operating beneath them.

In spite of these advantages, Gage's reliance upon meta-analysis and other recombinative procedures remains questionable. The use of meta-analysis involves the researcher in a number of new difficulties concerned with the nature of the relationship between the primary studies and the analysis carried out in combining them. For example, when one combines results of two studies, done on different populations with different statistical procedures, are the results strictly comparable? Are we sure that the phenomena from two different studies are defined in similar enough ways? Dare we accept a strengthened finding from results that were themselves weak? Criticisms of this nature were considered by Glass and his colleagues (1981), and we shall not review them here. An additional comment needs to be raised, though. Combining results in the ways suggested by Glass treats each original study as a single data point in a different combination of data. Glass proposes only one step up from the original level, but in principle there is no reason for stopping at the first metalevel. One could carry out the procedure *ad infinitum*, ending with a final analysis (Dillon, 1982). Faust would have been pleased with such a procedure, perhaps - but what he would have achieved at the end of the line wouldn't be knowledge at all, for at each step of the meta-analysis, more empirical content is lost. That is, we are no longer talking about the subject of investigation when we compare different studies of the subject. Rather, we now have a new study, with data that are different in kind from the original. At each step, we get further and further from the world.

Whether the loss of "empirical content" is justifiable in order to be able to compare studies in such a manner is something that will differ from case to case. What needs to be noticed is that no new explanations and no new (predictions of) facts come in with the technique. And insofar as the criteria for progressive science approach Lakatos's (1970) suggestion that a theory must explain or predict new facts if it is to be taken to be progressive, then meta-analytic techniques must be viewed as a reinterpretation of old things at best, and at worst as a merely defensive move.

If productivity were the only issue, we could not reach a firm conclusion that the process-product paradigm has reached the end of its usefulness. The slight and inconsistent results can be reinterpreted and explained away - Gage suggests that one should not expect anything better, given the size of the usual sample and given the interference of other factors (1978, pp. 26-27). What can be said is that the lack of more significant results suggests that there may be weaknesses in other places in the tradition that will explain what the problem is and perhaps suggest other ways of looking at the phenomena of teaching.

Conceptual (and Metaphysical) Anomalies

Empirical problems are only one side of the assessment of research traditions. As Laudan (1977) has shown, conceptual issues form an important part of a tradition's stock of problems. The critic needs to look in two directions here: internally to the theory or the tradition and externally to the relations between the tradition and other scientific and nonscientific theories. Conceptual problems have wide-ranging ramifications, for an inconsistency within a theory or tradition may carry throughout the theory; and contradiction between a scientific theory and other theories (or the encompassing world views expressed in common sense) may point to a problem on either side. Not all such conflicts are as one-sided as Galileo's conflict with the church in the seventeenth century.

Furthermore, internal conceptual problems may reflect more deep-seated problems with the basic ontology of a tradition - that is, conceptual problems may be a reflection of decisions made in establishing the theoretical "hard core" that postulate entities that turn out to be chimeras or that turn out to misrepresent the reality that they are trying to encompass, characterize, or describe.

Perhaps the most central conceptual problems for Gage stem from his definitions of teaching, when these are brought up against (1) his own comments on the place of learning in theories of teaching, (2) some tighter analyses of the concept found in the psychological and philosophical literature, and (3) the uses to which the concept is to be put in the research tradition itself and in the classroom. The last point overlaps with normative issues that come up in the research tradition.

Although Gage's definitions of 'teaching' are related, there are significant differences. "By teaching, we mean, for the present purpose of defining research on teaching, any interpersonal influence aimed at changing the ways in which other persons can or will behave" (1963a, p. 96). "By *teaching* I mean any activity on the part of one person intended to facilitate learning on the part of another" (1978, p. 14). The first definition carefully avoids using the term 'learning,' substituting for it a behaviorist euphemism, "changing behavior." Although Gage had explicitly attempted in that paper to separate (if not divorce) teaching from theories of learning, he here lets a behaviorist theory slip in through his definition. Behaviorist theories of learning emphasized the observable behavior that in other theories was used as *evidence* that learning has taken place. In part, for the behaviorist and for Gage, this is a methodological convenience without ontological overtones; if one deals in one's scientific investigations solely with the directly observable "behavior," one need not solve the mysteries of the mind. For a research program like Gage's, the advantages should be obvious: where one is concerned to correlate two variables, anything that enables one to avoid complications intervening factors is to be cheered. The "low-inference" character of behavior is another feature in its favor. But it should not be supposed that Gage has thereby avoided one thing he set out to sidestep, namely, the place of learning theory in teaching theory. For in this (and in the second definition) learning has a definitional relationship with teaching, as the thing to be achieved by the teacher's activities. One does not - indeed cannot - "read off" activities for teachers from learning theory, but one must have some theory of learning, however primitive, in order to put a theory of teaching into action or into research.

Whether one takes a behaviorist theory of learning or a more "cognitive" theory, then, might have considerable effect upon the nature of the theory of teaching that results. In Gage's case, the connection with the then dominant theory of learning and methodology in investigation assumed an answer that fit neatly into the fundamental methodology of his paradigm, the analysis of measurable variables.

More serious, perhaps, is the fact that both of Gage's definitions require an intentional view of teaching. In both, teaching is seen as *aimed at or intended* to bring about some effect on the student. In itself, this is impeccably correct; it is the activities of people who intend to bring about learning that are at stake here, not the activities of those from whom one might learn something incidentally, but who did not intend that such learning take place. Furthermore, Gage comes down squarely on the side of those who refuse to define teaching as actually causing learning - teaching is not to be defined as success at its own intentions. The students' achievements - the dependent variables in his analyses - are to been seen as the success criteria of an activity otherwise defined or identified.

What is questionable about such definitions in this context is that the methodology for studying teaching ignores the intentions of the agent. Rather than showing a direct concern for the intentional interactions of teachers and students, Gage's methodology could discover only more or less mechanistic relationships between teacher activities and student achievement. The intentions are turned over to "art," perhaps, but they are not reflected in the research itself.

One critic of process-product research has emphasized this point, particularly as it relates to the translation of the findings of process-product research into directives for teacher education. Fenstermacher (1979) contends that these researchers convert their correlations directly into rules for teachers' conduct and hand them down to the teachers without regard for the ways in which those rules fit into the teachers' own "subjectively reasonable" beliefs about teaching. The unwary reader could easily misconstrue Fenstermacher's arguments as being directed toward the logic of the conversion of research findings into practical rules, as if there were a fallacy within the practical syllogism. But the force of his argument is in fact pedagogical. It is a poor way of improving teaching because it fails to respect the subjectively reasonable beliefs of prospective teachers. Let us expand on this point, for some serious logical or methodological problems are hinted at by Fenstermacher's approach.

Fenstermacher presents the "logic" of conversion from research findings to teacher rules as a set of answers given to three questions.

> Q1: Do teacher performances P1 and P2 result in success at tasks K1 by students assigned to this task? [Researcher's answer: yes, by the correlations found.]
>
> Q2: Why do P1 and P2 result in student success at K1? [Answer: because P's probably cause K's.]
>
> Q3: What should teachers do in order to be effective in getting students to succeed at K1 and tasks like it? [Answer: P's.] (1979, pp. 163-168)

He notes that the usual procedure is to ignore the theoretical necessities of Q2 and in effect to move directly from the discovered correlations to the direct answer to Q3.

> This sequence of reasoning from Q1 through Q2 to Q3 is a kind of triple play. By answering one question, Q1, all questions are answered simultaneously. By coupling a few assumptions and presuppositions with a knowing wink at the absence of explanatory theory, all three questions get knocked off the playing field in the haste to move from modest correlational findings to imperatives for teacher training. (1979, p. 165)

And what Fenstermacher fears is that such a move will lead to a "basic skills" view of teaching: teachers will have to learn to follow the rules that are the answers to Q3 without attention to how they fit together with the preresearch beliefs.

Fenstermacher here points to a problem about the social sciences that philosophers such as Lakatos and Laudan have not handled adequately - the normative component that seems essential to social science, while being at most peripheral to natural science. It has become almost a truism that social sciences are "policy sciences." Fenstermacher quotes Ryan (1970): "The social sciences are preeminently 'policy sciences,'...they have been developed by and for men who have wanted to use the knowledge they could gain to bring about changes of one or another kind." The social sciences, including educational research, are application oriented - they are directed toward practical goals not just incidentally, but essentially.

The presence of a normative component in social research makes the logic of such research - or its application - somewhat different from that of research in the natural sciences. Traditionally this difference has been seen as one or another modification of Aristotle's practical syllogism. In a practical syllogism, the minor premise states what is possible. Such premises are the direct result of scientific (or other) research. The major premise states what goals ought to be pursued - they express values. A tension here between is and ought can be ferreted out along the lines first suggested by Hume, but this is not our task.

Process-product research follows the social sciences as a whole in having a normative component. As noted above, Gage criticizes scholars like B. O. Smith who seem merely to give a "naturalistic description" of teaching. (Descriptions are assumed to avoid value issues altogether.) Gage finds mere description inadequate. "I want to study teaching for the purpose of improving learning" (1966, p. 35). Improvement, of course, is a patently normative concept. Although improvement is not necessary for process-product research, it lies at the heart of its practical justification. The incorporation of normative considerations into his research tradition leaves Gage open to criticisms that he would not otherwise be exposed to. Kenneth A. Strike expresses the difficulty this way: "Situations do not become problems unless we approach them with values which specify what properties these situations ought to have" (1979, p. 10).

But let us take a closer look at how the findings of process-product research might fit into a practical syllogism intended to bring about improvement in teaching. Our major premise will be:

> 1. We want to improve teaching, to make it more effective.

A minor premise, garnered from research studies, would be:

> 2. Teacher's actions of type P lead to results of type K (a criterion of effective teaching).

The conclusion would be:

> 3. Get teachers to do actions of type P.

One can imagine all the qualifiers that go along with this: P must be the only method we know to achieve K, or P is more efficient and effective than some other known methods; K is accepted as an adequate criterion, etc. Other complex interaction effects must be taken into consideration also, such as how the research enters the wider intentional contexts of teacher education and school practice.

There is nothing invalid about this example of practical syllogisms using the types of research that Gage proposes as normal science in education. The problem arises out of "the fact that the school situation is made up of persons who act intentionally within a complex social system" (Fenstermacher, 1979, p. 159). Unless one is attuned to such a complex belief system, it is easily possible to misinterpret an isolated research finding as guide to action. This possibility is exacerbated when the research is conducted in one language, that of variable analysis, and is to be applied in another, the intentional language of action and belief. Considerations such as these give rise to Fenstermacher's concern about who determines what effective teaching amounts to - who plays the role of science (i.e., who determines what can be done) and who plays the role of art (i.e., who determines first what ought to be done and, perhaps more important, how to interpret the findings and convert them into principles for teacher education).

Gage seems to handle this issue by an appeal to complexity: science can handle the simple questions, but art has to deal with the complex.

> Scientific method can contribute relationships between variables taken two at a time and even in the form of interactions, three or perhaps four or more at a time. Beyond say, four, the usefulness of what science can give the teacher begins to weaken, because teachers cannot apply, at least not without help and not on the run, the more complex interactions. At this point, the teacher as artist must step in and make clinical, or artistic judgments about the best way to teach. (1978, p. 20)

But he is not clear how the teacher is to use the findings of research - he himself proposes a set of what he calls "teacher-should statements" that are cast as low-order rules for classroom organization and conduct. Two samples: "Teachers should have a system of rules that allow pupils to attend to their personal and procedural needs *without* 'drill' type" (1978, p. 39). These rules are presented as the result of "inference" from process-product research.

> We carefully sifted in this way [i.e., through meta-analysis, comparative studies, etc.] the detailed information for several hundred variables in teacher behavior and classroom activity. From this sifting we developed a set of inferences as to how third-grade teachers should work if they wish to maximize achievement in reading and, we think, also in mathematics, for children either higher or lower in academic orientation. (1978, p. 38)

But neither the sense in which these are inferences or the rules for assessing and making such inferences are clear here or elsewhere in Gage's work. The rules seem to be read directly off the research findings without concern for the teacher's preresearch beliefs, thus opening Gage up to exactly the kind of concern raised by Fenstermacher.

The problem with the practical syllogism noted above lies not with its validity, but rather with misinterpretations that may arise in converting from the premises to conclusions for teacher education. The syllogism may be valid, even sound; but the conclusion may not be adequate within the full intentional-educational context. What we should be trying to find out is not only that certain teacher actions lead to certain kinds of results (often, usually, or whatever), but also how these results may be brought to bear in complex intentional contexts where, among other things, there is the possibility of abuse.

Perhaps the difficulties of translating research findings into practical uses are not insuperable; conferences on the use of research in teaching are annual events in educational circles. What is not usually a concern of such conferences is the attempt to show that the difficulties are with the research tradition itself. And the difficulties with the research tradition are themselves very deep. Various writers have pointed to the depths: Alan R. Tom (1980), for example, comes to the conclusion that research cannot help solve problems of teacher education, because the normative/intentional nature of education is not amenable to empirical investigation. His radically pessimistic conclusion is that traditional modes of empirical research have no place in the training of teachers. At the core of Tom's reasoning is his assertion that teaching is an intentional, goal-directed activity. This introduces the concept of purpose into the arena of educational research. Because, Tom reasons, educational phenomena are purposeful or, to be more specific, because educational phenomena are subject to human purposes, such phenomena are unstable (1980, pp. 20-21). This position will be referred to as Tom's *instability thesis*. This thesis arises from the telic character of the idea of purpose. Purposeful phenomena change as human goals change, for example, as public policy and institutions are established, evolve, and are eventually abandoned. Tom's thesis cuts deep in that it challenges the very capability of research in education to satisfy even the rudimentary metaphysical requirements of *any* natural science.

Tom addresses himself to a number of the metaphysical prerequisites of science enumerated earlier in this paper. If the instability thesis is correct, then we cannot assume the uniformity of nature. If this is so, we must immediately abandon the principle of causality (1980, p. 23). We must further despair of ever arriving at the empirical generalizations required to establish natural laws (1980, pp. 18, 19, 27). In fine, we must abandon nomological theory altogether. Tom is explicit in declaring that the problems he poses are not merely "technical," that he intends "to throw doubt on the idea that the fundamental problems of education and teacher education are soluble only if we develop an empirically based instructional theory" (1980, pp. 26-27).

A great deal of the substance of Tom's observations concerning both the factual content and theoretical well-foundedness of process-product research is also found in the critical self-reflection of Lee J. Cronbach (1975). Reflecting on his own research into Aptitude-Treatment-Interactions (ATI's), Cronbach comes to pessimistic conclusions about the rate of theoretical progress and about inconsistent findings from similar inquiries (1975, pp. 116, 119). He attributes these failures to the ever-present "higher-order interactions."

> An ATI result can be taken as a general conclusion only if it is not in turn moderated by further variables. If Aptitude x Treatment x Sex interact, for example, then the Aptitude x Treatment effect does not tell the story. Once we attend to interactions, we enter a hall of mirrors that extends to infinity. However far we carry our analysis to third order or fifth order or any order - untested interactions of a still higher order can be envisioned. (1975, p. 119)

These interaction effects lead Cronbach to question the possibility of general theory in social science and education; generalizations are possible, but they are historically and geographically contingent - i.e., dependent on and limited to particular times and places.

In the face of this, Cronbach, unlike Tom, remains a guarded optimist. One important point registered by Cronbach against Tom is that "man and his creations are part of the natural world" (1975, p. 123). Since it is a natural phenomenon, human behavior is subject to natural laws. If these are contingent on time and geography, they are laws all the same. When we state generalizations in educational research, it may be necessary to provide qualifying clauses. The social scientist must be even more careful than the physical scientist to "state the boundary conditions that limit its [the law's] application" (1975, p. 123). In his own research, Cronbach emphasizes "intensive local observation," and, presumably, localized theorizing.

Whether the above caveats can satisfy those who, like Patrick Suppes, seek "theoretical palaces," Cronbach doubts; he bravely faces the theoretical and empirical facts without loss of faith. His example serves to remind us that the ideal model of science offered by the natural sciences can be rejected without abandoning science in its entirety. It may well turn out that such ideal notions of the natural sciences as the "free from external influence system" may be exposed as just that - ideals - useful only in suitably restricted domains. Such recognition has already taken place in microphysics, and Cronbach is right to extend this awareness to educational research. His paraphrase of Hamlet is well taken: "There are more things in heaven and earth than are dreamt of in our hypotheses..." (1975, p. 124).

And there are uses for social science as well, but they seem different in tone from Gage's: "Systematic inquiry can realistically hope to make two contributions. One reasonable aspiration is to assess local events accurately, to improve short-term control. The other reasonable aspiration is to develop explanatory concepts, concepts that will help people use their heads" (1975, p. 126).

More recently, two critics, David P. Ericson and Frederick S. Ellett, Jr., have drawn conclusions similar to Tom and Cronbach, although from a different perspective. After reviewing Cronbach's dilemma, they conclude that there are only two alternatives for social-science research: "(1) the wholesale rejection of the empiricist program in educational and social science research ... or (2) the attempt to reconstitute the empiricist science of human behavior on different conceptual foundations" (Ericson and Ellett, 1982, p. 506). Finding several attempts at the latter task wanting, Ericson and Ellett suggest that it is not mere lack of success at the tasks of science that is the cause. Rather, in principle, the empiricist program cannot hope to get at the significant matters of social life, where meaning and intention are the crucial concepts; this, they contend, always involves interpretation, and interpretation cannot be replaced by the sort of causal investigation that is at the heart of empiricist research. This leads to a rejection of the empiricist program, which is to be replaced by a model of understanding that differs from it.

> Our coin of knowledge is not firm generalizations, but is more akin to the good measure of meanings: plausibility. In educational research, as in education as a whole, good judgment should be seen as the prized intellectual capacity. Good judgment will not yield certainty, but it can yield interpretations and analyses far more acute and powerful than even the most skillful application of the empiricist "scientific-method." (Ericson and Ellett, 1982, p. 511)

The force of criticisms such as these should not be underemphasized. What is under attack is not a specific theory, nor a set of findings about the world, but rather the fundamental assumptions about the tasks of empirical inquiry. The very possibility of the world remaining stable, the viability of laws discovered in the search for causes of learning, and even the possibility of objective, rigorous investigation are thrown into doubt. It is small wonder, then, that a committed researcher like Gage will find ways of rejecting such claims. For this is what he finally does.

Gage's Response

The last chapter of *The Scientific Basis of the Art of Teaching* (entitled "Improving What We Know: A Reexamination of Paradigms") is Gage's defense of the process-product paradigm against critics like these. The Cronbachian view that laws and generalizations in social sciences change as time goes on Gage had considered earlier, in a response to Gergen. "Such possible failures of behavioral laws to have 'transhistorical validity' need not concern students of teaching in any culture whose values are shared by the influencers and the influenced. . . .The results of research on teaching may not last eternally, but they should last long enough to have practical value in any given generation" (1978, p. 19). But the difficulties of this are to be handled by the teacher as artist.

> Thus we see that the relationships between teacher behavior variables and what students learn should be applied with due regard for other factors, such as the ability and anxiety level of the pupils, the objectives of teaching, the characteristics of various pupils in the class, or the specific task to be taught at any given moment. . . . The general relationships that teachers work with must be combined in complex ways as adaptations are made to the particulars of any specific problem. (1978, pp. 19-20)

In Gage's view, the lack of success of the research tradition in providing guidance to teachers can be attributed to complexity; Ericson and Ellett's comment is directly relevant here: "The complexity of social phenomena is not a viable excuse for the small progress of the social sciences" (Ericson and Ellett, 1982, p. 506).[3]

In his response to critics, Gage exhibits the characteristics of a sophisticated defender of a tradition: he shows how criticisms can be turned into support of his own program, how his own program can be expanded theoretically to handle difficult objections, and how other methods fit into his own. The first part of the chapter handles basic metaphysical or ontological problems (he calls them "substantive" issues as opposed to the later "methodological" ones).

Perhaps the most significant example handled by Gage as an ontological question is that presented by Doyle (1978). Doyle had expressed doubts about the adequacy of process-product research to explain what happens in teaching. Doyle's criticism had been fundamental: the two-variable analysis that forms the heart of the process-product tradition overlooked the activities of students in

3. Their arguments to this conclusion are too complex to summarize neatly; see pages 505-6 for their full treatment.

interacting with the materials being taught, skipping them in the search for a causal relationship between the teacher's activities and the students' achievement. His contention was that these "mediating processes" may tell the story in teaching, but that the process-product paradigm rules them out of consideration. Gage handles this criticism, interestingly, by suggesting that the process-product model be expanded or elaborated to include the mediating processes.

The effect of this move is to absorb the criticism by model articulation. Gage's process-product paradigm now includes a new type of variable, viz., the students' mental processes, as exhibited in such "responses" as "paying attention, getting interested, becoming active, using imagery and mnemonic devices, engaging in self-recitation, making responses, and attending to feedback" (Gage, 1978, p. 74). These in turn are seen as "eventuating" in the products - "the kinds of achievement and attitudes that we have set up as the objectives of what goes on in the classroom" (1978, p. 74). It should be noted that Gage's way of stating these "mediating processes" is cast in the observable-behavior mode so essential to empirical investigation of human conduct. In place of his original model of teacher-process/student achievement which looks like this:

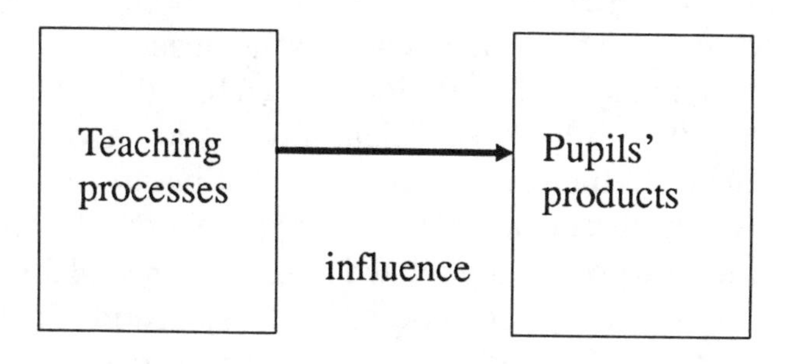

Figure 1.

Gage gives us this:

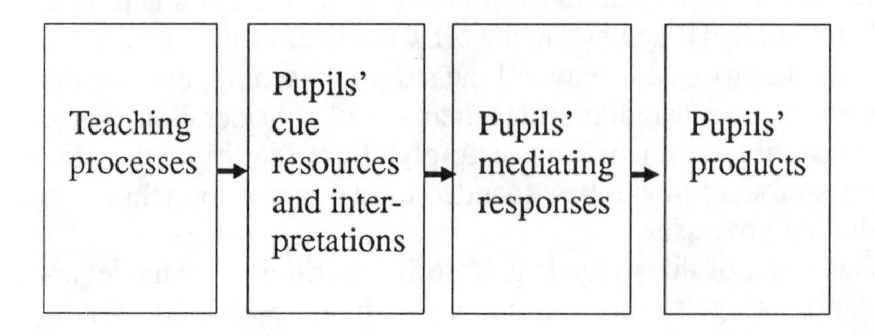

(Influence in direction of arrows)

Figure 2. (After Gage, 1978, p. 73)

Gage accuses Doyle of moving away from a concern with teaching, into a concern for "paradigms for research on any kind of determiner of student achievement" (1978, p. 73). Hence, the insertion of the mediating processes between teacher behavior and student achievement. What Gage seems to miss here is the possibility that teacher behavior is part of the mediating situation itself - not as a separate causal element, but as an essential part of teaching considered as a whole. It is not clear that Doyle sees this point, but the research he reviews seems to emphasize it. The teacher enters into this stage as well as into teaching as a separable item. "Just knowing the relation of a technique to terminal performance fails to supply sufficient information about immediate contingencies in the classroom. A mediating process interpretation, however, directs a teacher to experiment with other procedures which are potentially related to activating student attention" (Doyle, 1978, p. 174). Considered as an expansion of the process-product paradigm, Gage's move can be considered progressive, the rational thing to do. Rather than take Doyle's advice to put emphasis on different methods and types of analysis, Gage shows how that approach is compatible with his own tradition. If this procedure works, the theories of the process-product tradition would have more power to handle the findings of researchers in an explanatory fashion. What Gage seems unaware of, however, is that the type of entity proposed by Doyle (and other critics of his ilk) may be different in kind from the variables standardly considered in the tradition; the methods of investigation appropriate for measurable variables of behavior/achievement are not obviously adequate for considering "mediating processes" - and Gage does not provide us with an argument for seeing them as if they were similar enough for variable-analysis treatment.

Interestingly, Gage proposes investigation of teacher beliefs that could almost have been written by Fenstermacher. He proposes investigation into teachers' "implicit theory of teaching," which "takes the form of a hierarchically structured set of beliefs about the proper ends and means of teaching, the characteristics of students, the modes of learning, and the ways in which all of these interact to govern the teacher's behavior at any given moment" (1978, p. 80). Basically, these theories take the form of "simple maxims of teacher conduct," which Gage thinks could be codified and investigated with an eye to discovering connections between teachers' beliefs and classroom conduct. He does not take the final step into consideration of how one could use research findings to change these beliefs and the resulting conduct, but he expresses the faith that such studies could close the gap between "knowledge that" and "knowledge how" (1978, p. 81). Again, however, we find that there is no attempt to make the research findings or procedures deal directly with the intentional aspects of the relations between teachers and students; intentionality comes in from the outside, as it were.

The challenge of interpretation is not usually put as neatly as Ericson and Ellett put it, but Gage's response to recent calls for a switch to "qualitative" methods of research - notably by Rist (1977), but also by other writers - shows the direction that he would take. In the section of the last chapter on "methodological issues," Gage shows that he recognizes that such methodological recommendations challenge the fundamental assumptions of a natural-science approach to educational research. Following Viehoever (1976), Gage summarizes his view of the qualitative approach as follows.

> The qualitative methodologist, in addition to focusing intensively on single events or situations, emphasizes subjective phenomena, carries out undirected observation, prefers inductive strategies and heuristics, emphasizes the situational context, observes performance in the natural setting rather than the test situation, rejects sophisticated statistical analyses and randomized designs, emphasizes divergent rather than convergent realities, and uses popularized and affect-laden terminology. (1978, pp. 82-83)

In a surprising move, Gage does not reject such methods outright but rather seeks to find ways in which qualitative methods can fit into his general pattern of research. Qualitative approaches are seen as valuable heuristically, but not as an alternative to variable analysis. The two form separate but complementary methods. "The qualitative researcher can discover new phenomena and relationships or create new hypotheses. The quantitative researcher is better able to test, validate, or justify the hypotheses" (1978, p. 83). This is in part because qualitative methodology is useful only in discovering what is possible, whereas what he seeks is what is probable, and the latter can only be shown by "analysis of frequencies of occurrence in a sample of observations" (1978, p. 83).

This move by Gage can be viewed in two ways: first, as an attempt to admit some scientific value to a procedure that he views as essentially unscientific; or second, as an attempt to write interpretive or qualitative methods out of the scientific basis of teaching altogether. He clearly does not imagine that such challenges seriously affect his basic assumptions and methods of investigation. This move is parallel to his coopting of Doyle's mediating-process paradigm of teaching. The process-product tradition remains untouched by such objections, because the objections themselves are turned into opportunities for elaborating his own paradigm. The radical force of the objections is denied.

In the final section of his last chapter, Gage considers several ways in which "research can influence stalled movements," appealing to historical examples: the development of flight, the discovery of penicillin, and the development of women's rights (1978, pp. 91-94). In all cases, he suggests that continuing research has led to significant improvements despite the predictions of doomsayers. He uses this as a basis for his own optimism about research on teaching: past successes in scientific, technological, and social areas have come after plodding research and development within recognized scientific traditions. But history is a double-edged sword in such contexts. Consider for example that portion of nineteenth-century chemistry that proceeded outside of the Proutian atomistic hypothesis. First advanced in 1815, this hypothesis struggled throughout the century for recognition that it never received from many of the most prominent chemists of the time. Today, we recognize it as the cornerstone of modern theories of atomic structure, completely vindicated as early as Rutherford's school in physics and in chemistry by the success of the predictions of the periodic table devised by Mendeleyev. A comment on the traditions of nineteenth-century chemistry by Soddy (1932), himself a student of Rutherford, is telling.

> There is something surely akin to if not transcending tragedy in the fate that has overtaken the life work of that distinguished galaxy of nineteenth-century chemists, rightly revered by their contemporaries as representing the crown and perfection of accurate scientific measurement. Their

> hard won results, for the moment at least, appear as of little interest and significance as the determination of the average weight of a collection of bottles, some of them full and some of them more or less empty. (Soddy, 1932, p. 140)

Gage concludes optimistically that if process-product research on teaching is stalled at the moment, what is needed is further elaboration and more plodding work.

> At least for the foreseeable future, research on teaching will proceed as "normal science" (Kuhn, 1962); that is, investigators will follow the elaborated process-product paradigm and work on cleaning up an enormous number of details in the unfinished business of the field. New variables will be identified, invented, and measured. More ingenious ways of relating the variables, especially in complex causal patterns, will be devised and exploited. Better qualitative investigations, more comprehensive correlational studies, more intensive and single-case experiments, and more comprehensive path analyses will be performed. Better meta-analyses will bring together the results of the research in more valid and interpretable ways. (1978, pp. 93-94)

What is remarkable about this passage is that it suggests an openness to different methodologies and new "paradigms." We can be sure that if the problems of teaching are not empirically intractable in principle, some such continuation of research would lead to improvements. Gage's real call here is not to the process-product tradition, but rather to any sort of continuing research on teaching. With this one cannot disagree. But we do have qualms about the continuing defense of the process-product tradition, however elaborated and however much hard work goes into further studies built on the fundamental assumptions and methodologies of that tradition.

Summing Up

There can be no straightforward addition and subtraction of results in a discussion like this, as we hope we have made clear. The central issue is not whether process-product research is true or false, but rather whether it lives up to the expectations one might have for any progressive research tradition. In the end, the health of a research tradition is to be gauged by its success at answering its own questions, at solving its own problems, whether those be empirical or conceptual in nature. In the case of practice- or policy-oriented science, the issue is doubly complicated by the normative elements of the tradition and its political and practical contexts. As we noted early on, we cannot expect a knockout blow to drive a theory, paradigm, or tradition from the arena. But it is possible that a TKO could occur with enough jabs and uppercuts. We think that the foregoing discussion suggests or provides reasonable grounds for pursuing research on teaching in radically new ways. Several features stand out.

1. Gage's standard procedure when faced with criticism is not to show that the results of his research tradition meet the criticism, but rather either to co-opt the criticism or else to sidestep it, to argue that it is not really a problem for his tradition. This happens most notably in response to critics like Doyle and Rist, but it is an essential part of this discussion almost everywhere. A fully healthy tradition would not make such moves.

2. The rate of progress in the process-product tradition is a source of concern for all commentators, including Gage. No one feels, it seems, that the present state of research on teaching is where it should be, given the resources and manpower put into it in the last twenty years or so. It is conceivable that the process-product tradition will meet with the stunning success of some other research programs, but the question now is, What choices should be made for the immediate future? Considering the low rate of progress and the persistent conceptual problems, does this tradi-

tion have a right to its continuing domination of the limited financial and human resources spent on educational research?

Laudan (1977) suggests a way of viewing this question, of seeing what the choices amount to. Scientists (and by extension in our area, practitioners and policy makers) must make choices of two different kinds: the first is whether to accept the theories of a research tradition, and by doing so, the tradition itself, i.e., to treat them as if they were true for purposes of "certain experiments or practical actions" (1977, p. 108). The second is whether to pursue a particular research tradition or theory as a field of investigation. History is rife with examples of cases in which scientists pursued new traditions that were at the time not obviously acceptable in the first sense. Laudan suggests that reasonable choices can be made in both of these contexts, but that the criteria differ; in the context of acceptance, the criterion will be the chosen tradition's being a better problem solver than its rivals; in the context of pursuit, the criterion will be the chosen tradition's rate of progress in solving problems, i.e., that the new tradition has a higher rate of progress than its rivals.

It is an important feature of this method of assessment that a particular tradition have rivals: one does not evaluate a theory or research tradition in vacuo, but always against the example of others. The question here must be: what are the rivals for the process-product tradition? There can be only hints here, for many rivals have not yet defined themselves clearly as rivals of the process-product tradition, since they are growing out of it. One can mention cognitive psychology (e.g., Resnick, 1981), "Teaching as a Linguistic Process" (J.L. Green, 1983), and various more philosophical approaches suggested by authors like Fenstermacher (1977), T.F. Green (1971), and others (Komisar, 1968). Gage, of course, defends against these rivals in the ways noted above, but he often seems not to see how devastating their changes would be to his own tradition.

Our own attempt in the rest of this book is to show one way in which intentionality can become a central feature of a viable empirical approach to teaching.

3. The slow rate of progress and lack of empirically significant results of the process-product tradition can perhaps be explained by its endemic conceptual problems. These themselves can be traced to deeper "metaphysical" issues having to do with the nature of science, the objects of investigation, and so forth. As we have tried to show, Gage lets his ontology follow from his methodology: he starts with a method and finds the objects in the world that fit it. This in turn makes him unable to deal with the central problem noted by so many critics including ourselves - the intentionality of teaching. Although Gage's definition of teaching is intentional, his research tradition avoids intentionality and his co-opting moves often seem designed to cover up the lack of attention to intentionality. This leads to a problem of putting the findings of the tradition into practice - translating the findings from the nonintentional language into the language that dominates the practical settings of the classroom and the policy making sessions.

As far as we can see, there is no clear way that Gage or process-product research in general can directly handle the problem of intentionality in teaching. It is one of those issues that are sidestepped by calling upon the distinction between art and science, a distinction that is open to question in many ways, both as an analytic distinction and as a distinction of role in science/policy institutions. It is exacerbated by a conception of teaching that seems almost hopelessly mechanical and without depth.

4. Finally, there is an underlying problem of causality that runs throughout research on teaching. Gage urges experimentation to establish causal laws of teaching to replace the correlations that the tradition has uncovered. But the notion of causation in teaching is itself poorly understood, we fear. Insofar as teaching is in-

tentional, or the relations between teachers and students involve beliefs, emotions, or other mental matters, the causal model will have to be radically different from the one that Gage seems to assume.

For Gage, causality has the status of an unquestioned and unquestionable first principle. In Laudanian terms, what this provides is a constraint upon the types of theories that can be developed within the domain covered by science. Gage's views on causation are at once too narrow, thus putting a tight constraint on methodology and findings, and too broad, giving no criteria for assessment. They are too narrow in that the methods used to determine or discover causes are too narrowly inductive, based upon a picture of scientific investigation that draws its conclusions only after the data are in, without an encompassing picture of what is happening. They are too broad in that no distinction seems to be made between different types of investigation that must go on in research on teaching. The following paragraph shows the problem.

> A picture begins to emerge: Teaching that leads optimally to knowledge and understanding of complex tasks or materials - as in reading, mathematics, chemistry, or French - must meet intricate requirements (which have been or might be revealed by research on the learning and teaching of such subjects) relative to the proper sequencing and elaboration of concepts, instances, principles and problems. (Gage 1978, p. 77)

The determination of the proper sequencing of materials, concepts, and problems does not on the face of it seem to be a matter for the sort of investigation that could be carried out under the process-product paradigm, though. For what will determine such sequences is a complex epistemological, logical, and psychological interaction of students' knowledge, teachers' beliefs about that knowledge, and the nature of the subject matter itself. What may look like a causal analysis - one which would, we suspect, lead to suspiciously high correlations among variables - actually involves a rather extended logical analysis. And this suggests a different notion of causation for use in the study of teaching.

Consider Gage's appeal to research on cigarette smoking and lung cancer (1978, pp. 20-21). The correlations between the two variables in this case, as Gage points out, are low - in the range of .14, hardly out of line with the correlations found between teacher behaviors and student achievement. We base important personal and political policies on such findings, on the assumption that we can conclude that smoking causes lung cancer. Why can't we do the same for teaching with conclusions like, "teacher directedness causes higher student achievement than teacher indirectedness"? Here we need a closer look at the very notion of causation and at the purposes of causal investigations - the explanation of the world.

One can defend the claim that cigarette smoking causes lung cancer despite the low correlations, because one can imagine the causal chain that would show how the ingestion of nicotine and tars leads to the development of tumors. The correlations do not provide an explanation of the observations they summarize, they provide a description of a phenomenon that needs causal explanation. We know that the full explanation is going to involve neurological and biological theories that speak directly of the bodily mechanisms involved: a story must be told, one that has a beginning, an end, recognizable characters, and a coherent plot. Statistical research has a place in this investigation, of course: if there are no statistically relevant findings for a causal story, doubt is cast upon the truth of the story itself. But the correlations do not tell the tale: they can be used only to suggest high points in the plot or to show the tale to be false.

Cancer research has moved beyond the search for correlations like these into a search for the mechanisms that explain them.

Such studies are unremittingly and unquestionably scientific, but they turn away from the correlations to look at something else.

And it is this search for the connecting story that we find lacking in most process-product research. The aforementioned stabs at rival traditions - cognitive psychology, etc. - are making steps in this direction, but we suspect that only when they look directly at teaching as an intentional interaction among human beings will a full-fledged rival tradition to process-product research be developed and presented for assessment.

What should we look for, then? An adequate replacement for the process-product tradition will have to do several things: first, it will have to solve some of the unresolved conceptual anomalies noted above (in particular, we suspect, the nature of causation in teaching); second, it will have to show some promise (at least) of solving the basic empirical problem of the tradition: the lack of consistent and significant experimental results; and third, it will have to provide some way of explaining the successes of the tradition.

The last expectation is probably the most important. It is not that a new tradition should "subsume" the findings of the older tradition in a logically coherent fashion (an expectation to be found in some positivist philosophers' discussion of progress in science), but rather that the new tradition should provide a coherent way of showing why the earlier one got the results that it did. If such a "paradigm" can grow from present research on teaching, there is much hope for the future.

But what should not be done is to defend process-product research on teaching at all costs. Theorists in this field should be constantly aware of the force of the criticisms that are more and more stridently being leveled against the basic assumptions, methods, and findings of process-product research. Defense is not necessarily progress.

In this book, we seek to develop an alternative way of conceptualizing and investigating teaching. We believe that this alternative can solve many of the problems of the old research tradition - especially those centering around intentionality - while offering the promise of future research progress. It has the added advantage of casting the language of research in the same model - intentional - as that of the classroom, thus bypassing many of the "translation" problems noted by critics of the process-product tradition of research on teaching.

CHAPTER IV

EROTETIC LOGIC AND TEACHING

How does erotetic logic help us see what is happening in teaching? That is the topic of this chapter. Answering this question will take us into some relatively formal notions of questions and answers, for it is there that the rigor of the approach can be seen most clearly.

We shall use a version of the Hintikka/Åqvist theory of erotetic logic throughout this chapter and the rest of the work; there are several reasons for this selection. First, and perhaps most important, we are more familiar with and personally comfortable with this logic than with others. Second, this logic is particularly adaptable to pedagogical contexts, with its emphasis upon the questioner's epistemic alternatives, or knowledge of the situation being questioned. Hintikkian erotetic logic is thoroughly intentional and hence adapts well to the intentional contexts of teaching. Third, this logic pays particular attention to - and has a well-developed criterion of - answerhood, of the logical relations between questions and answers. Such a criterion is necessary if we are to be able to provide a criterion of completeness in teaching of the sort that seems to be necessary. Finally, the emphasis upon the imperative operator, "Bring it about that I know. . ." in Hintikkian logic ties up well with a new analysis of causation in intentional contexts - a feature of the erotetic approach to teaching that seems particularly important. This will be the focus of the next chapter.

We shall begin with a very simple example, in order to show the structure of questions and answers; later we shall show how different forms of questions call for different criteria of completion and conclusiveness.

Imagine Albert returning to his hometown in Middle America. A new house stands on Main Street. He asks Bertha,

(1) Who lives in that house?

From the question, we gain a great deal of information about Albert. For example, we normally can infer the following:

i. That he does not know the answer to the question.
ii. That he believes that there is an answer to the question.
iii. That he wants to know the answer to the question.
iv. That he believes that Bertha does know the answer or could provide an answer.

Note that we say "normally can infer". There are recognizable circumstances under which any or all of these may be false. All, for example, if it is a play we're watching and Albert is an actor (*not* a character) performing a role. The first, when Albert is a teacher giving a test (more on such cases later). The second when Albert is joking: he knows it's the new city hall and is intentionally putting down its opulence or its lack of character. The third when he's just filling the air with conversational noise - keeping the channels open, as it were. The fourth, perhaps, when the point of the "question" (note the scare-quotes) is to express amazement - "Wow! Who lives there?!" There are undoubtedly other cases in which these implications might be denied. Suffice it to say that it is the denials that need explanation, not the affirmation: in the normal case, these hold and they set the nature of the conversation between Albert and Bertha.

This group of implications is perfectly general, holding for any form of question, who, what, why, when, how, and so forth. These are features of questioning in general, setting out the conditions of the game.

From Albert's question (1), we are also entitled to infer some specifics:

a. That he believes it is a house (and not a stage set, city hall, etc.).

b. That he believes that someone lives in the house.

c. That he believes that the answer will include mention of or reference to at least one individual.

Again, there might be exceptions to one or all of these conditions, but again, it is the exception that needs explanation, not the standard case.

The first set of implications justifies a translation of any question into an imperative form; the questioner desires the answer to change his or her state of knowledge. So Albert in effect is saying to Bertha

(2) Bring it about that I know who lives in that house.[1]

The second set of implications, on the other hand, unites Albert's current state of knowledge about the house with what he desires to come to know about it. This may be put as the last part of (2):

(3) I know who lives in that house.

1. The question may seem more gentle, but the imperative nature of questioning may better explain why in some cultures questioning is considered impolite (Goody, 1978). It may also explain why some questions are not politely put by young Americans to older ones.

This will be the focus of our attention for much of the following discussion, for here we can begin to show the logical relations between questions and answers. Throughout this discussion, the imperative operator, "Bring it about that . . ." must be understood; it will be considered later in the discussion of intentional causation.

Following Hintikka (1976), we shall call (3) the "desideratum" of the question, since what Albert desires in asking it is that he come to know who lives there. We shall tentatively use the following symbols to formalize the desideratum:

(4) Ex K_A (x lives in that house)

or even more generally, using *S* to stand for any sentence,

(5) Ex K_A (Sx).

A translation of this reads,

(6) There is at least one individual, x, such that A knows that sentence S is true of x

Note that by now Albert has disappeared and we are speaking very generally and formally of any individual for whom the letter A may stand in, and the question, "Who lives in that house?" similarly disappears into its form:

(7) ! Ex K_A (Sx)

where "!" stands for the imperative, "Bring it about that ..."

Another thing has happened; we can see more clearly the different types of presuppositions which led us to this formalization. The first group has become the imperative; the second has the logical form of (5), but without the epistemic operator, "K_A":

(8) Ex (Sx).

Thus, this move to formalization of the original question asked anecdotally by a particular individual has led us to a coldly formal, though precise logical formula. What has been gained by such a move?

First, we are in a position to compare otherwise dissimilar questions for their logical structure. "Who lives in that house?" is *prima facie* different from other wh- questions,[2] but the form of a mother's question,

(9) When did you come home?"

is the same as the who-question (7):

! Ex K_I (Sx)

only now x ranges over time periods or instants rather than individual persons. Thus, we are in a position to make extremely general remarks about apparently different questions.

Second, we can now consider the formal conditions or criteria of answerhood, for these are logical issues as much as they are "content" issues.

Third, the formalization will enable us later to deal with questions of strategies in teaching, and, when we get there, with causation in intentional contexts.

As new symbols and logical moves come into the discussion that follows, we shall try to define them precisely, but we shall not attempt to give a full lesson on formal logic in this context; for that purpose, the reader is referred to any good beginning text.

2. "Wh-questions" are those whose operators begin, in English, with the letters *wh*: who, what, when, where, why, and so forth; they include also how.

Answers to Questions

When is a response to a question to be counted as an answer? When as a complete answer? When as a conclusive answer?

Note first, that the response must be in the right categorial ballpark: "The numeral 3" seems outlandish as a response to Albert's question. This is not exactly a logical point, however. It relates to the values of x in the quantified presupposition,

(5)Ex (Sx).

Similarly, "The richest man in town" is inadequate in response to a question "What is the square-root of 9?" ("The numeral 3" is not a good answer to this either, but that's another story.)

Why is it that "49" is not a good answer to (1)? Not because there is *no* circumstance under which it could be relevant. Imagine the context: in Sopchoppy, the people all have the same name, and they call each other by numbers. "49" could be an answer if such were the case. In normal contexts - those in which such a story is not forthcoming - "Who" calls for a reference to a person. Asking "Who lives in that house?" suggests the questioner's limitation of the answer to persons; "What lives in that house?" opens up the possibilities of such things as termites and bats as well as people.

It is important to recognize here that the questions one asks are ways of expressing and limiting intentions. To answer a question is to satisfy the expectations and intentions of the questioner in the questioner's terms, but not all the intentions that the questioner might have for asking that question. Perhaps Albert wants to embarrass Bertha by asking that question. The interplay of the conventional intentions of questions (which can be discussed logically) and the questioner's extra-conventional intentions (which cannot) provides the complexity of social life, language games, etc. (Grice, 1975, shows the complexities of such things.)

To get at the nature of answers to questions, let us make one further breakdown of our question, to get the presupposition without the quantifier. The presupposition of (1), remember, is

(5) Ex (Sx)

The "matrix" of the question is nothing more than the presupposition without the quantifier, i.e.,

(10) (Sx)

With this in mind, we can say this: any true answer to question (1) will provide a true substitution instance for its matrix; for example

(11) Charles lives in that house.

or

(12) The richest man in town lives in that house.

Either of these, if true, could be a relevant answer to Albert's question. Note that both might be true - a feature of some importance to the question of conclusive and complete answerhood.

Logically, the problem is this: a true substitution instance of the matrix of the question in response to the question does not necessarily provide grounds for saying that the individual is the same one referred to by the variable x in the original formulation. The presupposition that someone lives in that house, remember, is logically represented without an epistemic operator. The presupposition alone is to be read *de re*, as presupposing something about the way the world is. What is brought about in the answer is not the existence of some person, x, but Albert's *knowing* that the person lives in that house. The person's existence is *given* if the presupposition is assumed true, and the question in this case is merely who x is.

Logically, what this means is that a true answer to a question, cast in the form

(13) K_A (Sb)

does not necessarily imply the desideratum. For example, suppose that Bertha's reply to Albert is (11) "Charles," but Albert does not know who Charles is. In order for the desideratum to be entirely satisfied, Albert must be able to say "I know who b (i.e., the individual referred to) is." To guarantee this it is necessary to introduce an additional premise in the answer

(14) (Ex) K_A (b=x).

This premise assures us that the answer "hooks up" with the questioner's background knowledge and purposes in asking the question, that the answer resolves the intellectual predicament that gave rise to the question in the first place. Logically the conjunction of (13) and (14) always implies the desideratum (5). A complete answer must satisfy the syntactical, semantical, and epistemological conditions of the desideratum. The first two conditions are met by any true answer to a question; i.e., (13) above. The third condition is met by (14). We might add that it is not necessary for "b" to be a complete answer that the questioner knows prior to the answer who "b" is, only that it is known *after* the answer is given, perhaps by means of additional information gathered through further related questions, answered perhaps by a long protracted lesson.

The most remarkable thing about the third condition of answerhood, (14), is that it involves us in extra-logical epistemological criteria, "pragmatics" in the logician's and linguist's sense of the term. Why? Because clearly "x" must be provided by the student's already existing background knowledge as qualified, perhaps, by the student's purpose in asking the question. The introduction of epistemological considerations into the solution of the problem of answerhood is especially important for pedagogical contexts where it helps to show how a student's background knowledge is logically related to pedagogical action.

With the addition of premise (14), we have a logically complete answer:

$$
\frac{\begin{array}{l} K_A\,(Sb) \\ K_A\,(b=x) \end{array}}{Ex\ K_A\,(Sx)}
$$

The importance of this move, as already indicated, is that A might not be able to connect *b* with someone he or she "knows" i.e., can name, describe, or perceptually recognize. Premise (14), and with it the notion of complete answerhood, is relative to the student's (A's) background knowledge at any given time. What is a complete answer on one occasion for the same person may not be on another. Likewise, what is an incomplete answer for one student is complete for another. Finally, what is an adequate answer for one purpose fails for another. Suppose for example that Bertha replies to Albert's question "Who lives there?" with

(15) The richest man in town

but Albert does not know to whom that description refers. (15) gives an answer to him; it delimits the field of possibilities for the identity of x, but it does not pin it down.

Albert might respond in one of two ways. He might reject the answer and simply repeat the question,

(16) That doesn't tell *me* who lives there (i.e., "But who lives there?")

or ask

(17) But who is the richest man in town?

(16) takes the answer to be incomplete--not a full answer. (17) takes it to be complete, but not adequate to resolve Albert's intellectual predicament, which (presumably) is to be able to put the individual who lives there into the context of his present knowledge and interest or to increase his store of knowledge by the amount of that one individual's identity.

Two possible interpretations of complete answerhood compete here. The first takes (16) for its model and holds that any answer that does not fully satisfy Albert's intellectual predicament is incomplete. On this reading, the complete answer only comes about when Albert is able to place the relevant person precisely in *his* world (or in the possible worlds that constitute his "epistemic alternatives"--the different scenarios he can conceive of). This approach threatens to set standards unreasonably high. One unfortunate consequence of accepting this standard is that it forbids the inquirer from using any information that may be gathered from an incomplete answer. Another is that it tends to shut off dialogue.

A better approach might model itself on (17) and take "The richest man in town" to be a complete answer in itself in that it puts a stop to Albert's asking "Who lives there?" But it recognizes that the question is part of a set of possible questions about the residents of the house, "The richest man in town" calls not for a repeat of "Who lives in that house?" but rather for the different question, "And who is that?". This second view suggests a notion of "chained" questions that together draw Albert even closer to the resolution of his original predicament: Who lives there, how is he to be identified, what are his important characteristics, and so forth. The answer "Charles lives there," after all, may call for "who is Charles?" which might then be answered, "The richest man in town." The advantages of this model over the previous one are twofold. It allows the inquirer to make use of the information obtained by the answer to help resolve the original intellectual predicament while tending to keep the dialogue going. Some of the possibilities here will become apparent in the next section on partial answers.

Actually, these two models need not be seen as competing at all, but rather as complementary. In order to see this, though, it is necessary to grasp a second relativity of complete answerhood. The same answer can be simultaneously an incomplete answer to one question, say, the original position specified by the desideratum, and a complete answer to another question related to the first. Incomplete answers that provide complete answers to related questions ought not to be rejected since they provide information that helps eliminate epistemic alternatives and allows the questioner to draw closer to the correct answer to the original question. This is the direction Hintikka takes (1976). Sometimes, as when the answer offered is already known to the questioner or is simply irrelevant, the appropriate response is simply to reject the answer and repeat the question. We choose to exploit this second relativity of answerhood in a somewhat different direction. Rather than speak of eliminating alternatives, a view that tends to emphasize the incompleteness of inconclusive answers, and involves us in a controversial semantics, we prefer to speak of inconclusive answers as complete answers to some relevant question that initiates and is a part of an extended "chain of questions" leading to a conclusive (complete) answer to the original question. This approach highlights the pragmatic premise (14) in a special way. In the course of a chain of questions leading to a conclusive answer, premise (14) must be reiterated many times, once for every complete answer to a smaller but related question. For example, a first answer, A_1, might teach us that x=a, a second, A_2, that x=b. The order in which these answers are obtained may or may not matter depending on the questioner's background knowledge. Also note that the conjunction of A_1 and A_2 could have been given as an extended answer (lesson) that would have provided a complete answer to the original question. Remember, it is not necessary that the students know prior to the answer who "b" is, only that they know *after* the answer is given. This presses the relativity in the opposite direction.

The double relativity of answerhood and our use of it here is inspired by a similar relativity in deductive proof. There, every line of the argument is simultaneously a conclusion in and of itself to another argument and a link in the chain of argumentation leading to the conclusion sought. Later, when we show how the answer to a question can replace a step of deductive inference we will have occasion to exploit this analogy further.

It is important for our purposes to emphasize that an answer is not absolute or abstract, but concrete and relative to the questioner's background knowledge and purpose for asking the question. It should not be concluded from this feature of questions and answers that all is subjective, if by that is meant that it is totally a function of the questioner's psychological states. For in questioning, the world enters in; the issue is the logical or formal relations between the questioner's state of mind and what he finds in his world or environment.

We have talked of the "epistemological state" of the questioner here; this is a very specific notion, relating to the student's knowledge of the subject matter in question. It is to be differentiated from more global notions such as might be found in psychological theories of mental development (e.g., Piagetian stages) or cognitive psychology, where generalizations about mental capacities, growth and powers might be made. These are relevant to teaching in different ways from the specific epistemological states, ways that we will consider in a later chapter.

Partial Answers

Not all responses or answers to questions give complete answers, when that is defined as satisfaction of the desideratum. A partial answer arises as in (13) when a true answer to the question does not imply the desideratum; that is when the questioner cannot truly say "I know who b is." This occurs when premise (14) does not hold. Partial answers are what was referred to earlier as inconclusive answers.

The desideratum of our ubiquitous question, "Who lives in that house?" is "I know who lives in that house." There are several ways in which an answer to the question could be only partial. These correspond to the ways in which "I know who lives in that house" would be filled in.

A complete answer should do one or more of the following:

1. point out the resident in a crowd,
2. name the resident,
3. state relevant characteristics of the resident.

Any combination of one or more of these may allow Albert to say truly "I know who b is," depending on his intellectual predicament. There is no intellectual predicament that would not be satisfied by some combination of the above.

A partial answer - one that does not satisfy premise (14) - forces the questioner, as we have already indicated, either to reject the answer and repeat the question or to accept the answer and continue the dialogue by posing another question. This dialogue is controlled by what we have called the "chain of questions." This in turn is controlled by the questioner's state of knowledge of the subject matter at the time. Although warranted on occasion, repeating exactly the same question usually leads nowhere. Let us see where the second response takes us.

We illustrate the second response by means of the following chain. For the sake of brevity and convenience each link represents a different kind of answer, one of each kind (1-3) above. In principle a chain may be of any length and use one or any combination of the three kinds of answer.

Imagine that it is a party. Albert asks Bertha:

Q_1: Who lives in that house?
A_1: That man [pointing].
Q_2: But who is that man?
A_2: Charles Carton.
Q_3: But who is Charles Carton?
A_3: The richest man in town; your second cousin once removed (etc.).

There are circumstances in which any one of these answers would be fully satisfactory depending on Albert's background knowledge and purpose or special interest in asking the question, for instance when Albert already knows the answer to the other. ("That one! That's Charles Carton, my second cousin once removed!") The circumstances in which any one of them can be seen as giving a complete answer, though, are those in which it completes the chain of questions involved in complete answerhood. The assessment of complete answerhood takes all together and terminates when Albert can truly say regarding some putative answer that premise (14) is true for him. The above "chain" actually forms a "circle" of complete answerhood; it makes no difference logically where one begins. The first link is relative to Albert's intellectual predicament.

Hence, there are circumstances in which any of these answers would also be a partial answer. An answer's being a partial answer to the question is a function of the questioner's intellectual predicament with regard to the subject being discussed. This is what leads us into the extralogical considerations specified by premise (14) - "I know who *b* is" (Ex K*A* (x=b)). But the ways in which the three types of answer can relate to the original question are quite different.

In order to show how each of these may be involved in arriving at a complete answer, we shall digress enough to make a few brief comments on each of these, which we shall label (1) "ostensive", (2) "naming" and (3) "descriptive" answers.

Ostensive answers involve pointing at a particular individual or otherwise showing something in the world to the questioner. Answers to other non wh-questions, like "how does this machine work?" or "How does one swim?" may involve demonstrations of great complexity, but they are in principle ostensive, like the introduction of Albert to Charles.

Logically, we need a slightly different symbol for quantification here. Where we have been using "Ex" to symbolize "There exists at least one individual x such that . . . ," we now reverse the E to get "$\exists x$"; the latter is a "perceptual" or "acquaintance" quantifier, as opposed to the former, which is used in descriptive contexts.

More importantly, we may need a wider range of epistemic operators to replace the knowledge-operator in the desideratum - *K*. Consider Albert at the party. He is introduced to Charles and can truthfully say "I know Charles." ($\exists x\ K_I\ (\exists x = \text{Charles})$) Now Dora appears and asks "Who here is Charles Carton?" presupposing that Charles is here, and desiring Albert to point him out to her. The desideratum here is that Dora see or perceive Charles. The question might be formulated

$$!\exists x\ P_D\ (C = x)$$

where P stands for "perceives that" or "sees that". The relevant presupposition is

(20) $\exists x\ (x = \text{Charles})$

and the desideratum

(21) $\exists x$ PD (x = Charles)

Note that a complete answer to Dora's ostensive question will tie up her background knowledge of Charles with her present perceptions. As an answer to "Who is Charles Carton?" or "Who lives here?" the ostensive answer is perhaps only part of the complete answer; it has to be connected up with other information.

Naming answers answer such questions as "Who is that?" (pointing) or "What is the name of the person who lives here?" The desideratum of such questions (which presuppose that the object or person referred to has a name) is that the name be made known to the questioner.

! ($\exists$x) ($\exists$y) K_A [(x is a person) & (y is a name) & (y is the name of x)]

The conditions of complete answerhood are appropriately more complex here, since what we have is a double quantifier in the desideratum. Complete answerhood will require that x be identified and that y be identified and the two be related. In principle, this is not different from our more general question. We need not go into the complexities of naming and names; we are not putting forth any theories about meaning as naming, or names as being with or without "real meaning" or any such thing. For our purposes, the proper name of an individual person will suffice as a model.

The information given for a more general question is again partial, for specifying the name of a person or object must again tie up with other areas in the questioner's background knowledge and special interests. It is a complete answer to "Who lives here?" only when the other questions in the chain that defines the complete answer are themselves already known or answered.

Descriptive answers give information about the otherwise specified individual in question. Here the purposes of the questioner come especially to the foreground, for what sort of descriptive information is desired is not given in general by a question like "Who lives in that house?" but rather by the questioner's purpose in asking. Albert, we have seen, may be interested in Charles's economic status or familial relationships, but Dora may be interested more in his church membership or moral goodness. This means that the questions which make up the chain for the complete answer will differ from one person to another.

This point can be put formally, in terms of the questioner's "intellectual predicaments". The relevant gaps in Albert's knowledge have to do with familiar familial relationships and financial status: among his chain of questions are things like

What relation is Charles Carton to me?

(which has the presupposition: "Charles Carton is related to me"), and

How does he rank financially in Hometown?

(which presupposes "Charles Carton has a financial status which can be specified.")

Dora's questions are related to moral status or social significance; she asks

What church does Charles Carton belong to?
(Presupposing that he belongs to a church,) or

Who is Charles Carton's wife?
(presupposing that he is married.)

The general point to be seen from these examples is that questioners ask their questions with background concerns that may differ from one person to another. The significance of this for teaching is not to be missed, for the lesson designed to answer one student's question will be different from another's, even when the subject seems to be the same.

The child who asks "Where did I come from?" may only want to know its hometown or birthplace, and it may have no interest in sexual or biological issues; indeed, at some stages of development, there may be no presuppositions about such things to form the basis of questions about its birth.

Differently, a child who asks "Why does the ocean go up and down?" may find "The moon does it" a satisfactory answer, however partial it seems to the physicist; perhaps all the child wanted to know was that it was not some person - perhaps mother - who made it happen.

Similarly, a grasp of the concept of mass required to calculate a trajectory in gunnery is far less discriminating and complete than that required for high energy physics. For one thing, it is not necessary to know that mass may be converted into energy in the first instance. Even so, a vague and inexact notion of mass is adequate for the purpose of sinking ships and destroying cities. Even a little bit of knowledge may yield (destructive) power.

When questions are viewed this way, we can see that the presupposition of a question is an essential part of the questioner's intellectual predicament. Indeed, it might almost be said to define the predicament, in that the presupposition shows the gap to be filled in answering the question - i.e., in teaching.

We can now put logical flesh on the pedagogical slogan that teaching is filling gaps in students' knowledge. Let us define the student's intellectual predicament as the "gap" in the student's logical space.

The physicist's question about the tides differs from the child's not in its form or surface content, but in what presuppositions and background knowledge lie behind it. "Why do the tides rise and fall?" asks either one, or perhaps "Why does the water have different levels at different times?" The physicist brings to the question a set of presuppositions or limits on the answer that make the simple "The moon does it" unsatisfactory for him. The child might, of course, accept this as a partial answer and know enough to ask "How does the moon do it?" since there is no obvious mechanism. Of such further questions are physicists made - through a long series of lessons by clever teachers.

The mature physicist's questions are more discriminating. They have more power to add new information and rule other information out of consideration. The child might have no reason for rejecting "invisible strings" as an answer to "How?" That is something that comes with further development of its knowledge.

What ought not (epistemologically and perhaps morally) be done is to answer the child's question as if its presuppositions were the same as the physicist's. This point hardly seems worth making, for everyone agrees that subject matter should be presented to children in a form appropriate to their age and interests.

But the erotetic approach, this shows, gives an explanation of why this cannot or should not be done: the child is not capable of seeing the relevance of the answer to a question for which it does not have the appropriate presuppositions. It's not some vague state of mind or knowledge that is referred to, but a very precise (in principle) statement of a presupposition to a question between the child's current epistemological state (background knowledge) and the knowledge state it is seeking to be brought into as specified by the desideratum. A complete answer will "bridge" this gap by specifying an answer that "hooks up" with the student's pre-existing background knowledge as expressed by premise (13). The intellectual predicament is also determined by the student's special interest or purpose. What is a complete answer in one context is incomplete in another. The effect of special interest is to highlight what regions of our background knowledge are deemed relevant to the question at hand. A complete answer must not only hook up with a student's background knowledge, but with that part of a student's background knowledge that is highlighted by special interest.

Premises (13) and (14) provide a sharp logical characterization of what it is to fill in or bridge the gap in knowledge. Likewise, the desideratum provides a logical characterization of the epistemological end product. Indeed, the only thing that remains shrouded in mystery is the student's background knowledge and special interest. Ultimately this is the domain of empirical psychology, but teacher's diagnostic questions (see chapters VII and VIII, below) can tell us a great deal of what teachers need to know in order to properly gauge the gap.

CHAPTER V

EROTETIC CAUSATION

While there has been considerable philosophical argument about the necessity of causal analyses in the more esoteric areas of physics and chemistry, and while there have even been suggestions that the notion of causation could and should be done away with in education and psychology, there is little doubt among most commentators that causation has a large place in education and its back-up sciences.

This hardly needs mentioning, since educators' primary concern is bringing about changes in the students in their charge. What does need mentioning is that there are problems with the concept of causation as it is found in educational research and discussion in general. In this chapter, we shall try first to specify a way of thinking about causation that is a bit different from most conceptions; second to sort out and criticize some of the elements of causation found in the work of one early educational researcher, E. L. Thorndike and his successors; and third, to spell out the elements involved in an erotetic concept of causation, one that takes full account of the intentional elements of teaching.

Thorndike is chosen for this discussion because his effect on educational research was profound, even to the point of setting the direction of educational research from its beginnings to the present. In addition to that fact, Thorndike is particularly important not only because he did significant research on learning and teaching, but also because he had the courage to talk about such research in what is a recognizably philosophical fashion, even though he professed ignorance of and lack of interest in philosophy (Joncich 1968, p. 64). He was concerned not only to find out how learning takes place, and how teaching makes a difference, but also to discuss the importance of his findings for practicing teachers and policy makers. In these discussions, Thorndike shows the typical concept of causation as it has been understood in educational research since that time. It will be worth while to examine his more general remarks by way of introduction to the topic.

Thorndike was a great pamphleteer for a scientific approach to teaching and education.

> The practical consequence of the fact that human nature and behavior are knowable, the same effect being always due to the same cause, should then be to encourage insight, experiment, and reason in man's dealings with himself. Scientific spirit and method will be rewarded in education as in the physical sciences. (1912; Joncich, p. 74)[1]

Note the metaphysics of that causal clause: "The same effect being always due to the same cause." Thorndike had no doubt that one could give causal descriptions and do analyses and experiments to discover the constant causes of specific kinds of human behavior, knowledge, attributes or whatever. He professes the faith of the scientist, that at least in his own realm of study every event has a cause, and that that cause is discoverable. This carries directly over into practical affairs:

> At the bottom of the endless variety of human nature and circumstance there are laws which act invariably and make possible the control of human education and progress by

1. All dated references are to Thorndike's work; page references are to Joncich (1962).

> reason. So the general rule of reason applies to education: *To produce a desired effect, find its cause and put that in action.* (1912, p. 73)

The point of research, of the search for causal laws, for Thorndike, was always to be found in the practical uses of the laws: once one knows them, one can call them into action for the manipulation of the situation for educational goals. Lest it be thought that Thorndike had a narrow version of education - one to fit his rather narrow view of educational causes (to be shown later) - the following quotation shows how broadly he conceived it.

> . . . the exact selection from the facts of a man's life which shall be called his education, may be decided by convenience. A thinker about human education may choose his subject matter freely from whatever sciences concern man. . . . No clear boundary separates man's education from the rest of his life. In the broadest sense his education *is* his life. (1912, pp. 71-72)

Causation as Narration

It is time now to take an excursion through the notion of causation; for we must have some framework in order to compare different concepts of causation. It is necessary to put the very notion itself into a broader context, one that enables comparisons to be made.

Here is one way of thinking of causation, a "narrative" thesis. A causal claim - whether general or specific - always tells (or implies) a story, a story of the connections of events and objects with one another. "He dipped the paddle, stroked just so, and the canoe moved forward." "She said, 'Newton's laws are relevant to the explanation of tidal flow,' and the students learned that they are. . . . " "This billiard ball hit that one at an angle of 45 degrees, and that one moved off in a reciprocal path." The causal claims are unstated in these sentences (the problem Hume tried to deal with, we suspect); but it is of such stories that causal claims are made up.

Let us reemphasize this: the central feature of causal claims is a story. It is not - *pace* most philosophers - a generalization which lies behind the story, or a set of probabilities, or a statistical correlation. The root notion of causation is all tied up with narration. Note that this means that the basic notion is *singular*, that specific events are explained by causal descriptions without a necessary appeal to general laws.

But scientists, as we all know, are searching for the general, not the specific, however much their methods may depend upon the specific. How does that feature of science fit this picture? The answer: what is sought in the search for causal laws is the *general* story of such events and objects, a narrative pattern that singular stories are to fall into. What is important in this is first, that the singular story is in some sense logically prior to the general story, and second that they are the same sort of thing, not qualitatively different.

What then, are the central decisions that scientists searching for causal laws must make? Note that we say "decisions", not "discoveries"; this is important, for the teller of the causal tale in some ways *decides* what tale he will tell; the element of *discovery* comes in later, when the investigator checks the verity of his tale. First there's the question of characters in the narrative: Who does what to whom? Note that in human affairs, there is a wide range of possibilities here: it might be one person doing something to another, it might be one idea bringing about another, or it might be a situational factor ("stimulus") bringing about a change in the pattern of behavior of an individual ("response"). How the investigator peoples his story is going to be central to the total investigation, to the types of laws that are developed and to the possibility of testing the verity of the story. The scientist, we might say, makes his "ontological commitments" at this point.

We shall not say much here about how investigators decide upon certain characters rather than others; much will depend upon their views of the world, upon the historical and intellectual milieu in which they find themselves. Much will depend upon previous investigators' ways of telling the tale; sometimes it will be a reaction to some implausible tale.[2]

There is something to be said for aesthetic criteria here, too.

Second, there's the plot, or the plot-form. How do these characters affect (and effect) one another? What sort of relationship will the investigator admit to be proper? Few of us nowadays will let God into singular or general causal stories, even if we are believers. The problem of "action at a distance" exercised early critics of Newton's theories of gravitation; that problem can be seen as a plot-problem, for the critics could not see how those characters interacted. It is a central part of the philosophical problem of mind-body relationships: how can a mental substance interact with a physical substance in a comprehensible way?

A central criterion of evaluation here - on the part both of investigators and of critics - will be the plausibility of the story: how the actors in the tale manage to play their roles. Outrageous plots have to be specially justified. What is not permissible is the non-plotted story, the equivalent of the novel with transferrable pages or endings. And again, there's an aesthetic criterion here: austerity of plot may be as important in science as it is in drama.

The third set of decisions have to do with *fit*: does the story really fit the situation in the world, or is it merely fiction? This is where science comes in, for the methods developed by scientists to assess the verity of causal tales are precise and rigorous. The ability of the scientist to test elements of the cast and plot against

2. Thorndike's criticism of Gestalt psychology involves this aspect from both sides: the Gestaltists were objecting to the characters in the connectionists' stories; but Thorndike found the Gestaltists' tales much too complicated.

the world is central, particularly in the case of general causal tales; one can see the development of statistical methods, experimental controls, and such like, as the attempt to be ever more careful in providing a warrant for the truth of the tales we tell.

Along with these decisions goes a whole background of methods and expectations; logic, for instance, is relevant always (nowhere is it allowed to have a character or an event which is logically impossible); sub-plots are developed by way of methodological improvements. For example, telescopes and microscopes involve considerable tales in themselves, but they serve a supportive role in most modern astronomy or biology. In addition, the accepted views of the prevalent culture are those against which investigators develop and test their tales: scientists speak to their lay peers as well as their scientific colleagues.

Changes in science will be seen in this picture as changes in one or another of these areas: what characters will be allowed? Witches once, but not now. What kind of story is appropriate? What twists of interaction can be shown to be in effect? The answer to such questions will differ from one age to another. It might well be that scientific revolutions can be seen as revolutions in narrative style as much as in the content of those narratives.[3]

This will suffice as a beginning on a narrative theory of causal explanation. It sets some questions to ask of a researcher like Thorndike, some ways of approaching changes in the style of educational research, some suggestions for possible new ways of telling the tale of teaching and learning - which is the game we are after.

3. C.f. Quine's (1951) comment on "posits": "Physical objects are conceptually imported into the situation as convenient intermediaries -- not by definition in terms of experience, but simply as irreducible posits comparable, epistemologically, to the gods of Homer. . . . [I]n point of epistemological footing the physical objects and the gods differ only in degree and not in kind. Both sorts of entities enter our conception only as cultural posits." (In Morick 1980, pp. 66-67.)

Thorndike's Tales

Let us ask of Thorndike on teaching and learning: What characters are central? What sort of plot is involved? How does one test the verity of the tales told?

Characters: There's an odd lot here. There are, of course, human beings, made "intelligent and useful and noble" by teachers acting "in accordance with the laws of the sciences of human nature" (1906, p. 60). There are of course "stimuli" (a word "used widely for any event which influences a person - for a word spoken to him, a look, a sentence which he reads, the air he breathes, etc., etc." (1906, p.60)). There are - of course again - "responses" ("any reaction made by him - a new thought, a feeling of interest, a bodily act, any mental or bodily condition resulting from the stimulus" (1906, p. 60)). There are habits, tendencies, instincts, and other such characters as well.

When he speaks about moral matters - about the importance of education and teaching, about the necessity of a scientific approach to teaching - Thorndike peoples his tale with recognizable characters in the pantheon of Western philosophy and psychology: individual people, ideas, thoughts, attitudes, and habits. When he speaks as a scientist, he speaks of stimuli and responses. This is, we suspect, because of the type of plot he is willing to develop or accept in his scientific experiments.

Plot: For Thorndike, the causal plot is a tale of mechanistic relations among lifeless elements. What is done in teaching is to set up "connections" between stimuli and responses. The teacher (who might be a parent) begins with students who have natural proclivities and ways of reacting and responding to the situations in which they find themselves. Using the laws of effect and exercise, the teacher brings about changes in the student by altering the environmental situation. The model for such stories is that of engineering, with a dose of agriculture thrown in.

Like Newton's principles in mechanics, the law of effect ("Satisfying results strengthen, and discomfort weakens, the bond between situation and response" (1912, p. 79)), and the law of exercise ("Other things being equal, exercise strengthens the bond between situation and response" (1912, p. 79)), serve to set the form of the plots of the causal tales told by Thorndike. They are the conditions of comprehensibility for the stories to be told in teaching-learning situations.

Tests: The basic test to be used is measurement and correlation of variables by statistical methods. This is too simple a characterization, for Thorndike was central to the development of methods of investigation in psychology and education, but this is the crucial method: find which elements in the situation are correlated with one another; alter the first to see if the second alters as well. The elements are the aforementioned stimuli and response, related mechanistically with one another, and the test is statistical and probabilistic.

This is a cartoon; Thorndike was too subtle in his methods of investigation to be as naive as this sounds. But we will stand by the basic claim about the stories he was willing to accept in reporting his investigations. The problem was to find the specific stimuli that would connect up with responses in a way that made sense in the mechanistic tale of learning that formed the basic plot of the story. The resulting habits of behavior form the final element. This story is put into a broader context.

> The work of teaching is to produce and prevent changes in human beings; to preserve and increase the desirable qualities of body, intellect and character and to get rid of the undesirable. To thus control human nature, the teacher needs to know it. To change what is into what ought to be, we need to know the laws by which the changes occur. (1906, p. 60)

The scientist finds these laws by those methods, shows them in that tale, and recommends that teachers follow the same plot line in their interactions with students.

The Search for New Narratives

Thorndike set the tone for educational research through much of the twentieth century. His importance - and the value of his contribution - must not be downplayed. But it is well worth asking whether the limits he put upon those causal narratives are worth keeping. The view of science that Thorndike borrowed from Newtonian mechanics may now be obsolete, even in the natural sciences. And the mechanistic view of human nature and learning that Thorndike used to such effect is open to serious challenge, both conceptual and empirical. Pragmatic considerations - including the inability of investigators following his general pattern of investigation and narration to come up with firm conclusions - also suggest a change. (See Chapter III above for a more extensive criticism of Thorndike's intellectual grandchildren.)

Thorndike's tales were important in their time. But it is more important now to try new plots and stories, different causal elements. That's the problem for current educational research to attend to.

What are the appropriate characters in the educational tale? That's the central question. Teachers and others concerned with everyday life deal with real people, not with responses; they deal with specific things like numbers, historical stories, sentence structures, foreign words, and paper dolls, not with stimuli. They act intentionally - with all that that entails about hopes and dreams, beliefs and values.

Can these be brought into a "scientific" approach to causation in education? Well, why not? Aren't these part of human nature?

Does that mean that they are beyond scientific investigation? Surely not.

In what follows we propose to rewrite the characterization, the types of interaction and the plot line for educational research. At the heart of this new style will lie a genre of narration that departs from that championed by Thorndike and his positivist followers[4] as the modern novel departs from the Medieval Romances.

Our basic approach is to give a *logical* description of the interaction of teacher, student and subject-matter. The structure of the story is the structure of question and answer. All else in teaching is peripheral, sub-plots necessary to the tale. The laws of effect and exercise (and others also) are relevant, but they are not the most important element in explanation and practice.

A New Narrative: Erotetic Causation

We will now begin to develop a new kind of narrative compatible with the earlier analysis of teaching and the logic of questions and answers displayed in the preceding chapters. This narrative results from piecing together the narrative components already assembled. We will give a more detailed logical analysis later, but for now we will provide merely the barest outline of characterization, plot and criteria of validation and fit. So once again, only this time in a new style, let us consider the constituents of the story.

The characters and plots of erotetic narratives emerge directly from the articles of the intentionalist manifesto (see Chapter 1, above). First is characterization: "People believe things about the world" (article I), and "People's beliefs change" (article II). Few

4. That Thorndike was within the general school of positivists is beyond question; he seemed not to use that term, but later discussions of his work emphasize this aspect of his thought. What is important are his beliefs about the nature of scientific research, not the label. See Joncich (1968), a biography of Thorndike entitled *The Sane Positivist.*

educators would care to debate these although many researchers deliberately choose to ignore beliefs and knowledge in their narrations. Beliefs, especially students' beliefs, are central characters in the erotetic narrative. Articles III and IV declare that "Changes in beliefs can be explained," and "The explanations of these changes in belief are causal explanations." Article V asserts that "reasons in the guise of intellectual, logical or teaching acts may function as causes." This means that rational teaching acts must be included among the characters of erotetic narration. The erotetic concept of teaching is, broadly speaking, rationalistic. Article VI asserts that "teaching acts, functioning as rational causes, cannot be of a sort that does not involve the meaning (semantical content) of the beliefs." This indicates that subject matter, the content of teaching, cannot be ignored in teaching. To students' beliefs and teachers' acts must be added the semantical content of the beliefs or subject matter. These characters find their way into the narrative in the guise of questions and their answers.

The plot of erotetic narration is made up of the interactions of students' beliefs, teachers' rational acts and the subject matter. The basic unit of plot is "the intention of teaching acts - to answer the questions that the auditor (student) epistemologically ought to ask, given his or her intellectual predicaments with regard to the subject matter." This basic unit may be linked with others like it to form an indefinite (if not infinite) number of plot lines or question-answer chains of inexhaustible complexity by using the principles of erotetic logic. These intentional plot lines are far removed from the quasi-mechanical relations of the positivistic narratives.

The measurement and correlation of variables by statistical methods as a means of testing the fit or validity of a narrative is unavailable within the confines of the erotetic narrative. How then are we to assess the validity of an erotetic story? The answer comes in two parts and is simple and straightforward. With regard to the basic ingredient of plot, that is, questions and answers about the subject matter, fit is determined logically by how well the answers semantically satisfy the desideratum of the question. Complete answers fit the question perfectly. In the case of extended plot lines the test of fit is even simpler and better known. Extended question and answer series resemble lines of reasoning - logical deduction. The test of acceptability for deductive reasoning is logical validity. A second test is logical soundness: are the original assumptions (about, for example, the students' background knowledge or beliefs) and the answers that follow in the course of developing the plot, true? Note that both of these tests are internal to the narrative. As we shall see later, this does not mean that there are no important empirical problems and tests to be made concerning the external worth of specific erotetic causal narratives.

What remains to be shown is how the narrative constituents of character, plot and test of fit can be drawn together to yield an erotetic notion of causation within the confines of the Hintikka/Åqvist version of erotetic logic previously introduced; it is to this task we now turn.

The Logic of Erotetic Narration

The Hintikka/Åqvist version of erotetic logic introduced above exhibits all the elements of the erotetic narrative style. All that remains to be done is to draw them together into a causal narrative style.

Recall the conditions of satisfaction for the desideratum of a question. The first or syntactical condition is purely formal: it refers to the logical form of the questions formulated in first-order logic - the desideratum and its derivability from a complete answer. (See premises 13 and 14 of Chapter IV.) The second or semantical condition refers to the fact that the answer is a (preferably true) substitution instance of the matrix (Premise 13). It is this semantic property that allows erotetic logic so readily to accommodate the content of teaching - the subject matter. This is enough to take erotetic logic beyond ordinary formal logics. The third or epistemological condition takes us further still, for it requires that we employ the results of epistemic logic. A satisfactory answer must satisfy the questioner and not just the question, since a satisfactory answer must provide a true substitution instance of a substitution value known to the questioner. This pragmatic principle, expressed by premise 14 of Chapter IV, goes far beyond other logics and makes this version of erotetic logic ideal for the study of teaching. Yet in spite of all these remarkable properties there is something else about this version of erotetic logic more remarkable still.

The Hintikka/Åqvist analysis depends on breaking questions down into two parts, "an imperative or optative operator plus a description of the cognitive situation the questioner wants brought about" (Hintikka 1976, p. 22). As we have already seen, this means that a question like

(1) Who lives in that house?

becomes

(2) Bring it about that I know who lives in that house.

where "Bring it about that" is the imperative or optative component.

It seems to us that the analysis of the structure of questions as consisting of an imperative (or, as we shall see, causal) operator plus a desideratum, is generally correct and fruitful. The role and behavior of the desideratum have already been extensively discussed above; here we consider more carefully the role of the imperative. Hintikka, primarily concerned only with arriving at a logical and semantical characterization of questions, intentionally overlooks the problems connected with the imperative or optative element (1976, p. 23). We, on the other hand, are seeking a logico-causal characterization of questions and the answers that bring about or cause their satisfaction, a characterization adequate to serve as a foundation for research on teaching. As a result we cannot overlook the role of the imperative in questions.

Earlier, we too ignored the optative by glossing over it with a symbol, '!', and avoided any further discussion of it in focusing on the structure of the desideratum. In this chapter things change considerably, with the optative moving to center stage. Remarkably, we can render (2) (awkwardly but accurately) as

(3) Cause it that I know who lives in that house.

This is the simple consequent of juxtapositioning the Hintikka/Åqvist analysis of questions with any view that sees "bringing something about" as a straightforwardly causal notion. (See, for example, Ennis 1983, and Gasking 1955.)

Although clumsy in normal English, this construction does no harm to the underlying logical-semantical structure. It has one important advantage over (2) above: by expressing the imperative in causal terms we obtain the completely general case. (We can express this generality by writing "Cause_____" where the stem '_____' may be left blank or take on a very broad range of substitutions as the situation requires. So expressed, the imperative may not only be treated generally, but is rendered flexible enough to handle a broad range of natural language questions.) But it is not the nomenclature of erotetic logic that provides the structure of an erotetic narrative; rather it is the special sense of connectedness contained within it that suffices to establish a logico-causal nexus.

The first and most obvious feature of erotetic narration is that it does not bring in an external relation between distinct empirical events. It is an internal logical relation between question and answer. The logical relation is not, however, one of implication; it is semantical and epistemological rather than merely syntactical, since content and comprehension are crucial. The logical relation and *ipso facto* the logico-causal nexus is the relation of *satisfaction*. Minimally, the answer must satisfy the content conditions specified by the matrix of the question. More fully it must satisfy the epistemological conditions specified by the epistemic operator (K_I, B_I, P_I, etc.). Stated differently, the answer must bring about the satisfaction of the content and epistemological conditions of the question. The relation is material, not merely formal. Satisfaction may here be taken as a relation between an open sentence (the matrix) and ordered n-tuples of objects, functions or relations that comprise the answer. Satisfaction, so understood, is a logical, semantical, and epistemological relation; but the satisfaction of questions must always occur within the scope of a causal (imperative) operator.

Two Narrative Styles

We have proposed an alternative to the type of narrative used by Thorndike and his empirically oriented followers in educational research. This alternative narrative takes the characters to be things like beliefs and knowledge, and questions and answers, the plot to be logical in form. Now the question arises: is the (erotetic) logical narrative merely a parallel description of the same events as a Thorndike or Gage might be concerned with, or is it a substitute?

This is an important question. If the two narratives merely provide alternate descriptions of the same event, they can co-exist without contradiction. Resulting research would throw light on different "facets" of the object under consideration. If, on the other hand, the two narratives provide contradictory stories, one type of research has to be seen as missing the essential nature of the object under consideration.

Underlying this issue is an important point about the description of the world: any object or event in the world can be correctly described in an indefinite (if not infinite) number of ways. How one chooses to describe the phenomena will be determined in large part by what purposes are to be served by the description. Different purposes lead to different descriptions. And the question now becomes, are the purposes of a Thorndikean description different from those of an erotetic description? For if they are, then the two descriptions may co-exist as parallel narratives.

But we think that the two compete; both attempt to give a description of how the activities of teachers cause changes in the students. Strangely, the logical picture provided in the erotetic narrative provides a strong causal picture of those changes, an "intentional" causal picture of the causation of changes in belief.

An Excursion into "Learning"

What needs to be explained here is a change in an individual's cognitive repertoire, specifically in what that individual believes about some aspect of the world or his/her place in it. More generally, what needs explaining is the individual's learning. A full theory of learning, viewed in a way totally compatible with the erotetic theory of teaching, has yet to be developed; but such a full theory is not necessary to our point. Only a sketch is possible now.

Conceive of learning not as an addition to a set of discrete beliefs (or even "representations", to borrow a term from cognitive psychology (Fodor 1981)), but as a change in the structure of the individual's mind or belief system. I.e., no individual holds beliefs in total isolation from other beliefs (Green 1968, 1971). And there will be holes in that belief system, places that the individual cannot fill from the current set of beliefs. The "holes" are logically defined, perhaps as questions that cannot be answered without going outside current knowledge or beliefs. The problem is, how does a particular hole get filled?

Consider an example. Bertha, we might assume, knows that there are different makes of cars, most of which she can recognize on sight; among these are Volvos and Toyotas. But she does not know (or for that matter, have any relevant beliefs about) whether one of these makes is larger than another. So a question is there: "Which is larger, a Volvo or a Toyota?" Walking down a street, she notices a Volvo parked next to a Toyota; the former is larger, a fact which she notices, and she answers her own question (which might not have been explicitly asked), "Volvos are larger than Toyotas." It seems apropos to describe this as Bertha's learning that Volvos are larger than Toyotas. (Note that she might have learned something that is not universally true: the generalization from a single case to the universal is fraught with difficulties. But without single cases, even with possible counter-examples, there is no learning.)

The important point is that Bertha's learning is described here "erotetically", as an answer to her (implicit or explicit) question. The learning is a dynamic relation between that question and the "fact" that provides, for her, the answer.

It's a large jump from a single example to holding that all learning can be so described; for the moment, we shall take it for granted, however, that that conclusion is appropriate. In part, we hold this because we cannot think of an example of learning - even skill learning (see Chapter II above and Macmillan and Garrison 1986) - that couldn't be so construed. Insofar as this is an appropriate way of thinking of learning, there is a foothold for seeing teaching and learning related as (erotetic) cause and effect.

Note that there are background conditions behind any causal description of learning; we could say of Bertha that the "experience" caused her learning that Volvos are larger than Toyotas. And that experience (i.e., seeing the two parked next to one another) requires that she have been attentive to the situation, that she have been aware of the two cars, etc., etc. These are preconditions of her learning, though; they are not the learning itself.

In a similar way, that question might have been answered by a teacher. Knowing that Bertha has that background knowledge, and also knowing about the gap in her knowledge, Albert might say (or demonstrate) something like "Volvos are larger than Toyotas." His assertion becomes the cause of her learning, but only in the context where those background conditions are met.

And Albert's statement causes her learning in a logical way. I.e., the relation between Albert's statement and Bertha's future state of knowledge is a logical relation every bit as much as it is an "empirical" one. If we want to find out what caused her learning, her coming to believe that Volvos are larger than Toyotas, all we need to know is what was said under what epistemological circumstances (or, perhaps, what "experiences" she had that led to that *as an answer to the question*).

Note further that what is important in this causal picture is the *content* of the interchange between Albert and Bertha. The tone of voice used, the pat on the back, or the psychic depth of the relationship between them may be important to Bertha's accepting the answer, but without the description involving Bertha's prior intellectual predicament, there is nothing to talk about. Any research into teaching which ignores such things just misses the point. *The central element in causal studies of teaching is the content of the interaction, seen as a logical relationship between teacher and learner.*

Back to Teaching as Causation

We have found that the erotetic notion of causality demands that we reject two major assumptions of the traditional analysis of causal plots: first, that causality is solely a relation between empirical events and second, that causal connection is distinct from logical connection. We are compensated for this loss by obtaining an erotetic notion of causality capable of rigorously capturing important aspects of the domain of educational research that include intentional contexts. The purpose or intention of the answer is to bring about the satisfaction of the content conditions of the question. The question, or more precisely, the desideratum of the question, may be viewed as the intentional object of the intentional act insofar as it specifies the success conditions for the intention.

Erotetic causality is an intentional notion of causality with the advantages of logical clarity and rigor that normally do not attend discussions of intentional matters. The overall advantages of an erotetic/intentional causality have already been hinted at; we will discuss more specific applications later. For now we would like to consider one of the most notable consequences of erotetic causality.

According to the traditional analysis of causality the causal nexus is not observable; it is argued that it is not possible to observe any necessary connection between events. It is possible to concede the latter half of this assumption while denying the first, at least insofar as erotetic causality is concerned. It may be shown logically that the answer provides a satisfactory substitution instance of the matrix of the question. Epistemologically, when the "I" of the epistemic operator (K_I of the question) is emphasized, the satisfaction of the question may be directly and immediately experienced, as when we exclaim "I understand" or "Now I see!"

This experience of satisfaction may be taken quite literally. To see this consider the case where the answer to the question *involves* showing the asker the answer. Suppose Dora asks "Who here is Charles Carton?" The answer may consist in merely pointing and encouraging Dora to follow the ostending finger to the spacio-temporal point containing Charles; if she recognizes the intended perceptual object, this will place her in direct perceptual acquaintance with the answer. Stated differently, Dora will experience the satisfaction of the question, or, stated in yet another way, Dora will experience the erotetic causal nexus. The pointing finger brings about the satisfaction of the content conditions of the question by placing the agent in immediate perceptual contact with the answer, thereby satisfying the desideratum of the question.

The same result could be achieved by using a description, but a picture or a perception being worth a thousand words, showing is almost always more satisfying - or at least more economical - when the proper satisfaction instance is immediately present. In general, descriptions are rendered superfluous on those occasions when it is possible to bring about immediate perceptual acquaintance.

What is experienced in erotetic causality is not a necessary connection between two distinct unrelated events; rather it is the presentation of a single logical event - a composite of question and answer. The logical event is the internal relation of satisfaction. Questions and answers are not distinct: the old adage, "A question determines its own answer," gets at the special sense of connectedness that holds here.

In erotetic causality the causal nexus is present and apparent to anyone who attends to the satisfaction of the content conditions of the question. It is not necessary to infer the causal relationship from an underlying regularity. In a sense every erotetic experiment is a crucial experiment. This does not mean that erotetic research that seeks to penetrate below the surface of discourse may not be revealing; but it does mean that the causal nexus is already apparent on the logical and semantical surface of things even if the full consequences of the connection lie deeper. In intentional contexts knowledge of the cause always brings some degree of immediate interpretive understanding although fuller understanding may require additional psychological, sociological and economic considerations. Much of such further research, however, lies outside the domain of any theory of teaching *per se*.

A related consequence of the transparency of the causal nexus is that the relation of satisfaction need not be generalizable. A singular causal statement need not entail a corresponding universal law. The composite question-answer relation may be a unique, one-time-only event.

There are important universal causal laws and we are particularly interested in the laws of classroom discourse. We believe it is the task of educational researchers to find them. But this belief does not commit us to the view that every particular causal relation must instantiate a universal law. Human intentional activity is often as unique and individual as fingerprints. Such events may not be predictable in a material sense, but they may nonetheless instantiate logically predetermined forms of discourse.

We have already mentioned some of the advantages of an erotetic notion of causality. Other advantages derive from the fact that it avoids many of the criticisms that plague the traditional analysis. For example, from the beginning, the Humean conclusion that we do not perceive the causal nexus has been constantly challenged. Critics have ranged from Hume's younger contemporary, Thomas Reid, to the contemporary Swiss psychologist Jean Piaget, including as well many other modern philosophers concerned with the philosophy of action (Kenny 1963, Davidson 1963, 1980, *et al*). We hope that erotetic causality will be seen as part of this common sense tradition.

Two reflections on erotetic causation deserve mention before we move on to some variations of style available to the erotetic teacher. First, it should be clear that we do not seek to eliminate all other narratives from the domain of research on teaching. We only wish to provide an alternative that is for the most part more appropriate to the domain of study - purposeful human conduct. Similar remarks hold for ethnographic and qualitative researchers looking at real episodes of teaching who expose many kinds of intentions, purposes and meanings hidden within the complexities of social contexts. We see no problems in developing many different narrative genres either to expose meanings or to reveal causes. It would be as foolish to say that there is only one correct or appropriate style for research on teaching as it would be to say that there is only one literary genre that is correct for depicting life itself. We do believe, however, that the narrative form of erotetic causation is vastly superior to most others for depicting the domain of human activity that concerns us here - teaching.

Second, and finally, it is interesting to note that as a literary or pedagogical genre, erotetic narrative is far from original even for the purposes we propose it to serve. The narrative we have been describing is closely related to the dialectic of the Platonic dialogues. To this basic style we have added the principles of erotetic logic and causation as well as an analysis of teaching as a question-answering activity. (For these reasons, we prefer to call our style dialogical rather than dialectical.)

The gist of this chapter has been that attention to the logical elements of the relationships among teacher, student and subject matter are central in any explanation of what happens in that relationship. Our contention is that those logical elements can best be seen as providing a causal explanation of what happens in teaching. The next chapters will investigate this claim, first by considering some general points about teaching strategies, and then by examining some examples of teaching.

CHAPTER VI

EROTETIC TEACHING STRATEGIES

In any given narrative style the choice of character and plot are unlimited. In this chapter we will explore some of the more obvious possibilities. Since it is our purpose to apply the erotetic narrative directly to the improvement of teaching we will refer to these narrative possibilities as "erotetic teaching strategies."

The precise criteria for successful teaching, hence the basis for its evaluation, are given by the three conditions of satisfactory answerhood. These conditions say nothing about *what* should be taught nor specifically *how* the satisfaction of the question is to be brought about. Selection of subject matter, hence the specific content of answers, lies beyond the scope of any theory of teaching *per se*, lying, as it does, in the realm of curriculum theory. On the other hand it is reasonable to expect a theory of teaching to address issues regarding how teaching might proceed. Erotetic theory treats problems regarding the planning and implementation of teaching; that is, the *process* of answering students' questions, as problems in selecting an appropriate question answering *strategy* from among the many logically equivalent alternatives. In this chapter we consider some of the purely formal strategic possibilities available to the teacher in answering students' questions. Determination of the best strategy is heavily dependent on context and can only be determined empirically.

It is unlikely that all of the logically equivalent strategies are in practice equally effective. Effective teaching strategy depends, as we shall see, on a great deal more than logic; nor is there any reason to assume that there is one best strategy for all occasions. The selection of an appropriate erotetic teaching strategy depends on a complex interaction of at least four things: erotetic logic, the structure of the subject matter, the individual characteristics of teacher and the characteristics of the students. There is a great deal of empirical work to be done here. Our immediate task is to set out some of the more interesting strategical alternatives along with a few definite constraints.

We start by distinguishing two different kinds of erotetic teaching strategies. The first is narrowly concerned with the order in which questions about subject matter are taken up and answers given. This order will be referred to as the "pedagogical order" and the accompanying strategic considerations as "pedagogical strategy."

The second kind of erotetic strategy deals with the logical dynamics characteristic of the interactive dialogue between student, teacher and subject matter. The concern is with the "moves" available to teacher and student and the "cost" of making them. We call this "dialogical strategy." Pedagogical and dialogical strategy are not entirely distinct. The former functions within the latter. Pedagogical and dialogical strategy will be taken up in that order.

Before beginning it is useful to note that the number of options available within either strategy is very large. Added to the fact that they interact in complex ways, the possibilities seem endless. This complexity may appear forbidding at first, but some important general guidelines quickly emerge. Besides, only this kind of logical complexity could hope to provide the sort of flexibility necessary if we are to come to grips with the uniqueness and the intricate intentionality of classroom discourse. Erotetic Theory and its strategy cannot and need not shy away from the intentionally complex, one-time-only, teaching encounter. It is not the case that anything goes in Erotetic Theory: there are strategic rules, guides and constraints. Nonetheless the horizon of possibilities does seem unlimited. What follows is only a glimpse of those horizons.

Pedagogical Strategy

Pedagogical strategy rests upon four ascending layers, each of which serves further to constrain the strategic possibilities left open by the layers that lie below. The four layers are deductive logic (which sets a deductive order to the lesson), the presuppositions of questions (which partially determine pedagogical order), the students' background knowledge (which sets the starting point for teaching), and finally teaching tactics (which deal particularly with the place of partial answers in teaching). At the very bottom rest the constraints of deductive logic; for example, the principle of noncontradiction, truth (or whatever) preserving rules of inference, etc; even such results of proof theory as the Gödel-Church theorem can be important. Deductive heuristics, e.g., the ancient method of analysis and synthesis and the results of proof theory, have a great deal to contribute to pedagogical strategy. To this may be added the heuristics and constraints of erotetic logic. For example, it is a universal law of erotetic heuristics, as applied to erotetic teaching, that everything else being equal, students ought to ask and have answered completely those questions that provide the most information regarding their intellectual predicaments. Needless to say the occasions when everything is not equal are legion. A second feature of the deductive heuristics of erotetic logic is truly remarkable. In carrying out deductions individual steps of inference may be replaced by answers to questions. The heuristic advantages are considerable. For example, suppose in the course of a proof it is necessary to determine S_1 v S_2 (possibly A v not A). If a suitable question can be asked and answered, a crucial experiment for instance, then our task is cut in half as compared to having to deal with both parts of the disjunct.

Pedagogically this means that there are frequently certain questions students epistemologically ought to ask themselves in order to "bring about" the right conclusion, or what amounts to the same thing, grasp the object of the lesson. Teachers' answers to student questions, even teachers' questions and pedagogical questions, may be quite helpful here.

It is fruitful to look upon the conclusion of a deductive argument as an answer to a question. This casts a new light on partial answers. If we consider the conclusion of an argument to be an answer to a question then it is quite natural to consider each line of the argument as part of the answer allowing us to draw closer to the complete answer. Those answers to "smaller" questions that allow us to carry out a single step toward answering the "big" question may be considered as partial answers. Later, in our discussion of dialogical strategy, there will be occasion to expand considerably on these introductory remarks emerging from the erotetic equivalence of deductive steps and interrogative steps in reasoning to a conclusion.

Like the heuristics of deduction the theory of proof must be augmented in the following way: in order for a question to be asked and answered it is necessary that all of its presuppositions first be established. Theoretically there can be no exceptions to this rule of erotetic logic although in pedagogical practice it is best to ease this requirement somewhat by allowing questions whose presuppositions are then provided by the teacher's answer. This will allow teachers to answer students' muddled but essentially appropriate explicit questions. This restriction on presuppositions determines a rule of pedagogical order that should never be broken; to do so involves one in the most tragic of erotetic fallacies. The logical error of asking a question before establishing its presuppositions has been known since antiquity as the fallacy of begging the question. This fallacy can lead to disastrous consequences; tragic reversals befell those who asked questions of the Oracle at Delphi for which they had not first carefully established the proper presuppositions. As a consequence they received dangerously ambiguous answers. If the Oracle had been more concerned with pedagogy than with prophecy she might have rejected ambiguous questions or clarified them, by providing the appropriate presuppositions either before answering or in the course of giving the answer.

A number of other classical fallacies arise as a result of confusion regarding the presuppositions of questions. For example the so called "fallacy of the complex question," exemplified by "Have you stopped abusing your students yet"? The "fallacy" here is the hidden and unestablished presupposition that you ever did abuse your students. Needless to say such erotetic fallacies should be avoided.

Determining which presuppositions to introduce, and in what order, allows the teacher, administrator, or curriculum planner to control both the content and direction of the teaching dialogue. All of the problems regarding pedagogical strategy come down to this: What presuppositions should the teacher instantiate and in what order? This is an extremely important and nontrivial question. Important because the presuppositions represent the content of the curriculum; logically nontrivial because even a Turing machine cannot always decide on what presuppositions to introduce in a logical proof so as to solve the problem, or answer the question. This is a direct result of the Gödel-Church theorem.[1] Restrictions of various kinds upon presuppositions govern

1. The introduction of auxiliary constructions in geometry is equivalent to introducing new individual constructs in first-order logic. Often times, the number of 'auxiliary' individuals required in the intermediate steps is greater than the number of individuals in either the theorem or the axioms. These larger numbers cannot always be predicted recursively on the basis of the theorem. In such cases the number of auxiliary individuals needed grows as a function of the Gödel number of the theorem faster than any function. For these there is no recursive decision procedure.

pedagogical strategy and are absolutely necessary for the success of any dialogical game.

Other restrictions on alternative strategies deserve mention. For one thing it is good strategy to provide true answers or at least answers as well warranted as the current state of human knowledge will allow. Teaching requires trust; the student must be confident in the veracity of the teacher's answers. Clearly subject matter competence is crucial here. Finally, teachers should reject any explicitly asked questions whose presuppositions are false and provide reasons for the rejection.

To the restrictions of formal logic must be added the constraints of content, i.e., the subject matter. Erotetic logic does not in any way specify content, (the curriculum planner does that). Whatever content is specified supplements the restrictions of erotetic logic with the internal structure of the subject matter. In any even loosely organized body of knowledge there are concepts that are related and structured by rules for their meaningful use. The rules and concepts are often so structured that it is logically impossible to acquire certain concepts before acquiring certain others. In classical Newtonian mechanics, for example, the concept of acceleration presupposes the concept of velocity just as the concept of force presupposes the concepts of mass and acceleration. Once acquired, the concept of force provides a general principle, a scientific law, that may be used as a premise from which to deduce further answers, i.e., to explain a large number of natural phenomena.

It is tempting to allow the logical order intrinsic to the subject matter to determine the pedagogical order. Erotetically this would amount to little more than ordering the presuppositions according to their logical dependence within the subject matter and carrying out the requisite derivations. Curriculum planning in such a case would be reduced to an exercise in deductive logic. What could be easier? Indeed, many have even claimed that all there is to teaching beyond subject matter competence is following the intrinsic logical order of the material. Some have even suggested that, since the latter is part of the former, subject matter competence is all there really is to teaching. This oversimplified view of teaching strategy is frequently championed by subject matter specialists against the claims of the educators. But one has only to consider the debacle of the "new math" to see where this kind of misguided strategy can lead. (See Kline 1966.)

It is easy to see what is wrong with simply converting logical order into pedagogical order. Hirst provides a useful analogy in this regard:

> Logical order is the end product in scientific understanding. It is a pattern which in teaching is pieced together as one puts together the pattern of a jig-saw. The logical order does not prescribe a series of steps which must be taken. There is a great variety of ways in which the jig-saw can be made up. (1979, p. 51)

Hirst is quick to note that it should not be concluded from this that "in developing understanding one can start *anywhere*" (p. 51). It is necessary to respect the logical order and structure of knowledge, but this only partially determines pedagogical order and strategy.

Erotetic theory clearly indicates what is wrong with an exclusively subject-matter oriented approach to pedagogy. It is impossible on the Erotetic Theory to ignore students and their individual differences. These differences involve students' individual purposes, interests and background knowledge including those presuppositions that determine what questions they are capable of asking. Those questions that the student epistemologically ought to ask must be drawn from this set of already existing presuppositions. These presuppositions determine where in the jig-saw it is possible to start because they tell us what pieces the student already has. Individual intellectual predicaments determine where it is best to begin. We should not forget though, that intellectual predicaments are *about* subject matter. This means that the logical arrangement of the subject matter cannot altogether be ignored. Finally the whole situation is considerably complicated, but not substantially altered, if the student has false presuppositions, pieces of the puzzle that don't belong. Students' intellectual predicaments determine where teaching must begin by determining what questions students epistemologically ought to ask first. The order in which a student ought to ask questions determines, in most cases, the best pedagogical order for that student. To establish the appropriate pedagogical order it is first necessary to determine the students' intellectual predicaments, i.e,. their special interest and background knowledge including what presuppositions, true or false, they already possess. It is this that makes teachers' diagnostic questions so important.

Pedagogical strategy is not entirely determined by the overlying layers of deductive logic, subject matter structure and students' intellectual predicaments. There is a fourth and final layer of pedagogical strategy determined by what we shall call teaching tactics, those options that remain open to the teacher or curriculum planner once the logical order has been determined by the first three layers. For example, even in a preconceived curriculum a number of questions may arise that the student epistemologically ought to ask. Similarly, which explicit questions to pursue, clarify or reject remains, in part, a tactical decision. Even the choice of whether or not to make the question the student ought to ask explicit is a tactical choice, a part of pedagogical strategy. Many issues regarding the best teaching tactics and consequently the best pedagogical strategy can once again only be decided empirically. In what follows we only seek to outline some of the tactical options open *a priori* to the teacher or curriculum planner.

Many tactical decisions revolve around the teacher's use of partial answers. Any lesson, and oftentimes an entire course of study, centers on one or a few questions that the teacher must answer. Depending on the size of the gap between the students' background knowledge and the epistemological state that must be brought about in order to satisfy the desideratum of the final question, it may be tactically advantageous to break the answer down into smaller steps of varying sizes, i.e., partial answers, in order to answer the final question completely. There are many logically equivalent ways of arriving at a complete answer to a given question. The situation is the same here as in our earlier observation that partial answers are steps in the derivation of the conclusion, the answer to some larger question.

As a rule, breaking the answer down into a chain of smaller questions has a number of advantages. Smaller questions are generally more compact and manageable for both student and teacher. This is especially important given the temporal constraints of the typical classroom.

In Chapter IV we saw that the completeness of an answer is relative in a number of different ways. One way of working this relativity to tactical advantage would be to make the partial answer come as a complete answer to a smaller question. The situation is not unlike the fact that any line in a deductive proof is simultaneously a conclusion, an end in itself, and a means to further ends, an intriguing unity of opposites for those such as John Dewey who delight in such things (Garrison, 1985). In any case this view of the function and use of partial answers opens up many important pedagogical applications.

One advantage of this approach is that it allows the teacher to exploit the natural fecundity of partial answers, their tendency to generate further questions forming a "chain of questions." Perhaps this is the pedagogical value of the so-called "pregnant question;" its complete answer spawns many lesser questions and answers. Partial answers give impetus to student-teacher dialogues. Each new partial answer, itself a complete answer to another question, produces a new intellectual predicament. This motivates (*logically*) another question that the student epistemologically ought to ask. Note that the sense of "motivation" used here is not in any way *psychological.* This same approach may contribute to holding or enhancing students' interest and motivating them psychologically. This might be achieved by making the intellectual predicament explicit and hoping that the student grasps the question, perhaps even framing it explicitly. On other occasions the teacher might find it efficacious to render the appropriate smaller question explicit, show its connection to the larger question, and then answer it. These are tactics open to the teacher. Although the latter group has a more psychological character, they all have a logical basis.

Another option emerges from taking the relativity of complete answerhood in the other direction. Complete answers are relative not only to the question but also to the students' background knowledge or purpose in asking it. Consequently the same answer may be complete for one student but not another. Recall the questions in Chapter IV where the child asks "Where did I come from?" or "Why does the ocean rise and fall?" One tactical option open to the teacher is to provide a more than complete answer, satisfy the desideratum of the question to the student's satisfaction and then provide additional information (perhaps even additional presuppositions). Sometimes students should not only be satisfied but over-satisfied. Students can sometimes grow into answers just as they grow into an older sibling's clothes. But this move can backfire if the answer overwhelms and confuses the student. One sure way for this to occur is when the answer extends not only beyond the students' background knowledge or purpose in asking the question, but their cognitive stage as well. This takes us into cognitive developmental psychology, a topic that lies beyond the scope of our theory. Nonetheless there are many points at which the psychological meets the logical. We consider some of these later.

Another tactical option involving partial answers is to provide a partial answer and leave the students to obtain the complete answer for themselves. This maneuver is likely to backfire unless the students have sufficient knowledge, a clear and explicit idea of the question to be answered, and well developed higher level cognitive skills that make them capable of acquiring and processing new as well as preexisting knowledge. Such a tactic could be positively disastrous should it turn out that the students lack the requisite presuppositions to complete the answer. It is most likely to succeed if the student already possesses the knowledge necessary to "hook up" with the partial answers provided. (An example is provided below in chapter VIII.) In effect this means that the answer given is actually complete if only the student "thinks about it," that is, arranges already existing knowledge in the appropriate way and/or elicits the requisite tacit knowledge. Teacher's pedagogical questions used in conjunction with partial answers can contribute a great deal to the successful use of this tactic. In general such mixed tactics could contribute to the teaching of higher order problem solving skills. It would probably be best when using this approach to make the question explicit.

Another variant to this tactical maneuver occurs when the teacher provides a partial answer in the form of an ellipsis wherein the student must fill in the omissions from background knowledge. Where to leave the "gaps" is a delicate tactical issue. In many ways this maneuver has more tactical advantages and fewer disadvantages than the related maneuver just discussed.

A separate set of strategical options revolve around the role of presuppositions in questioning and answering. Generally teachers ought to reject all explicitly asked questions whose presuppositions are false. Occasionally they may choose to accept such questions, but not before correcting any false presuppositions or restructuring the question into a related question all of whose presuppositions are true. On other occasions the teacher might even find it efficacious to provide additional presuppositions in order to arrive at an appropriate question.

It is tempting to say that no student ought to ask a question that has a false presupposition. But this would weaken the power of the erotetic conception. To see why, consider an example (first suggested by Max Black in a seminar in the early 1960's): A child asks, "Why is that cow giving white milk?" A serious answer to this question might go into the processes by which milk is made from digested grasses, the ways in which bovine biology works, and so forth. But if we know the background of that question (Black called it the "preamble"), "Cows give milk that is the same color as their skin, and that is a brown cow," we recognize that the presupposition is something like "That cow is giving the wrong color milk." The question *ought* to be asked, given that presupposition. The teacher, of course, ought not to answer the question directly, but deal with the presupposition instead. The point of teachers' diagnostic questions is not merely to discover what true presuppositions the students have in their cognitive repertoires, but also find out what false presuppositions must be dealt with.

The preceding remarks barely scratch the surface of the use of presuppositions in pedagogical strategy. For the most part the direction of question-answer dialogues is controlled by what presuppositions the teacher selects to introduce (instantiate) and in what order. Introducing presuppositions opens up new lines of inquiry whereas withholding presuppositions can shut a line of inquiry off, since a question simply should not be asked unless its presuppositions have already been established.

There is one very special limiting case of teacher-student dialogues where the requisite presuppositions are not introduced but rather are discovered. Here the possibilities are not only unlimited but unknown. For Plato dialectic, as exemplified by the Socratic dialogues, was the genuine inductive *science* of first principles (and presuppositions) and the means to their discovery. The logic of questions and answers was employed as an instrument for achieving new knowledge. In such dialogues the role of teacher and student could reverse, indeed the two roles often became indistinct. A question-answer dialogue that leads to the discovery of original presuppositions is of course an ideal limiting case, yet something of the thrill of the original discovery of a new presupposition or even the answers to entire questions ought to accompany every instance of first (original) learning.

Pedagogical strategy provides a useful framework for understanding and controlling the impetus and direction of teaching based on the logical structure of the subject matter and the student's intellectual predicament. This approach to teaching strategy tends to focus on and grow out of the logical structure of the subject matter. However, the quasi-deductive character of pedagogical strategy restricts its range of effective application. It provides, nonetheless, a foundation for looking at classroom teaching dialogues and their strategy in a rigorous logical manner. Pedagogical strategy functions within a broader and more inclusive domain, one that emphasizes classroom dialogue and the interaction between student, teacher and subject matter; that is, the domain of dialogical strategy.

Dialogical Strategy

In this section we turn to a discussion of dialogical strategy. Dialogical teaching strategy draws its philosophical inspiration from Wittgenstein's notion of language games; in this case the game is teaching. In his paper "A Dialogical Model of Teaching" Jaakko Hintikka (1982) has attempted to cash in Wittgenstein's notion using the mathematical theory of games. It is Hintikka's dialogical model that we follow, *mutatis mutandis*, below. We begin by specifying the structure of the "game."

In the basic game there are two players, the teacher (T) and the student (S), each of whom has a store of information (background knowledge) in the form of a list of assertive sentences. There is also a separate store of information, again in the form of assertive sentences, call it the subject matter (M). Conceived most broadly M could be understood as the store of all knowledge available within the culture at large. Initially it may be assumed that there is no overlap between M and S's lists while M and T's lists do overlap.

The following kinds of moves are possible:

> (i) T can make an assertive statement to S. This is called a teaching move. A teaching move is an intentional act of teaching directed toward answering the question S epistemologically ought to ask given his or her intellectual predicament. This move and these acts lie at the very core of teaching. All of the options of teaching tactics and pedagogical strategy are available to T in planning and implementing this move.
>
> (ii) T can specify a finite subset m of M. The sentences in m are then conveyed to S by some means other than move (i). This move is the counterpart to assigning homework, lab work, assigned reading, computer assisted instruction, and so on. This is called a "study move". We must be careful not to confuse the activity of answering questions (teaching) with such study "moves", lest we confuse the assigning of students to question answering sources with the activity of teaching *per se*. (See Ennis 1986, for such a confusion.) In this move the student obtains the answers to questions independently of T. The fact that study moves are distinct from teaching moves reflects this independence. Indeed, the role of the teacher may be eliminated completely in self study, the role of the classroom teacher anyway.

In moves (i) and (ii) there is only some varying degree of probability that the information conveyed to S will actually be encoded. This "noise" in moves (i) and (ii) means in effect that what is taught is not always learned. This is merely the dialogical concomitant of the fact that what is taught is not always learned. This reflects the truism that teaching is neither necessary or sufficient for learning.

(iii) T may address a question to S, who gives T as complete an answer as he or she can or else denies the presupposition of the question. These questions are what we will call teachers' diagnostic questions and pedagogical (information ordering and eliciting) questions. These questions are semantically different, at least upon the surface, from ordinary information seeking questions. Why? Simply because the teacher already, presumably, knows the answer to the question. Asking and answering such questions calls for more than is usually found within the confines of the language game of information seeking. Below the surface, such questions are information seeking. T seeks information about information; that is T seeks information about S's information state. This deeper structure does not appear in the semantics provided in Chapter IV, although some surface ripples do appear in the iterated intentional operator. To attempt to plumb the depths here would require semantical research that lies beyond the scope of our present effort. But this semantic peculiarity, as we shall see, has considerable strategical importance.

(iv) S can ask T an explicit question. T's reply is an instance of move (i), an act of teaching. This move is constrained by the presuppositions in S's background knowledge. T may exercise any number of the tactical options left open by pedagogical strategy; e.g., not to answer, to answer partially or completely, or to reject the presupposition of the question.

(v) S can perform a step of deduction from the information already acquired. The conclusion is then added to S's list of sentences. Teachers' questions can contribute heavily to the efficacy of this move. Even questions students pose to themselves may help here.

(vi) T can obtain information from M. These moves may be conveniently diagrammed for easy reference.

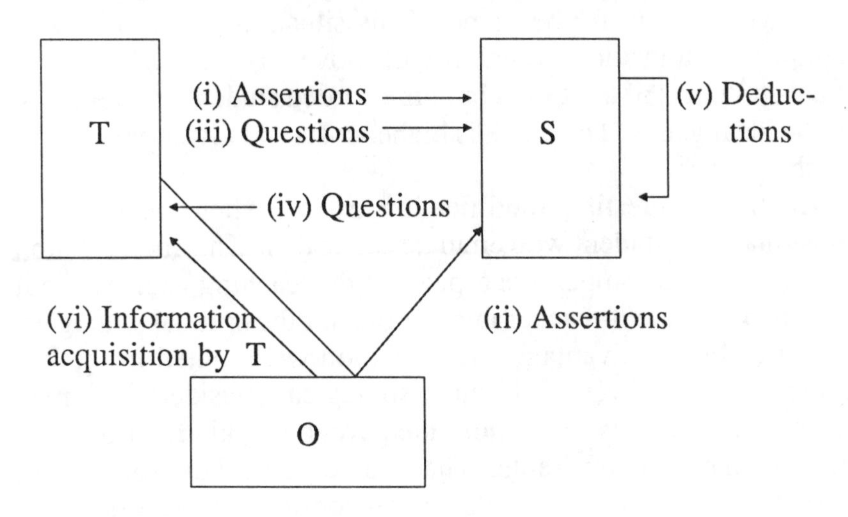

(*Figure 1. After Hintikka [1982], p. 42*)

Dialogical strategy is determined by the "costs" and "payoffs" of the various moves. Costs are assessed in terms of time and fiscal expense of materials. Optimal strategy seeks to minimize costs and maximize payoffs. So conceived, moves (ii) and (vi) would appear fairly expensive in time and material. In contrast moves (i) (iii) and (v) are generally inexpensive, as is (iv) as long as T can answer the question.

The payoffs for both T and S depend on the total information content of S's list at the end. In other words the payoffs for S and T are higher the more S learns. Consequently the dialogical game is a cooperative game: S and T have a common goal. This of course describes the ideal game into which other elements may enter to break down shared goals and introduce less cooperative play. This can occur by changing the payoff schedule, for example when the teacher is rewarded and the student is not, or at least does not perceive the reward. In such circumstances S will be discouraged from making use of the most effective strategic tools available. Normally these are moves of type (iv) and (v). S might even withdraw from the game. This situation is also likely to dampen the transition probability of moves (i) and (ii). The optimal strategies of a rational but unrewarded student are not those of the ideal game. The result is higher costs and lower payoffs for all the players.

Another interesting modification occurs when we consider more than one student who emulate each other. This modification emphasizes the competitive aspects of the teaching game without altering the student-teacher interaction, nor the optimal strategies. Historically the advantages of emulation were touted in the last century by Pestalozzi. Although strategical considerations may remain unaltered by emulation, the psychological effect of competition may be undesirable. That is an empirical question; these psychological effects lie outside the confines of the dialogical game *per se* There are other modifications that do not directly effect the cooperative nature of the teaching game. Some of these we will take up later in this chapter.

Having introduced them, let us now take a somewhat closer look at moves i-vi, and how they interact. Let us begin with move (i), teachers' answers to questions students epistemologically ought to ask, since these lie at the heart of erotetic teaching. The three criteria of answerhood given in Chapter IV set firm conditions for the satisfactory use of this move. The same criteria also provide conditions for the causal efficacy of this move. Furthermore most of the options within pedagogical strategy depend upon this move as their *modus operandi*. Effective use of pedagogical strategy greatly enhances the transmission probability for the information contained in this move. For all these reasons this move lies at the very core of the Erotetic Theory.

It is imperative to the whole dialogical game that T be able to answer S's questions, otherwise move (iv) becomes prohibitively expensive. At this point the entire game begins to collapse. The greater expense of move (vi) is justified only by what it eventually contributes to directly reducing the cost of move (i) and indirectly by means of the latter to move (iv). Move (i) lies at the center of the dialogical web.

Next come students' questions, move (iv). Answering these questions is the object of teaching. We may partition these questions this way. First there are the questions that students epistemologically ought to ask given their intellectual predicament. These questions may or may not be explicitly asked. Second we have those questions that *are* explicitly asked. These questions divide roughly into those that ought to be asked and those that should not. The latter include questions with false presuppositions and those intended to embarrass the teacher or another student. Like teachers' answers, students' questions have been studied extensively in Chapter IV. The causal conditions for their satisfaction have been stated in Chapter V. Very little remains to be said.

One interesting function of students' questions that we have so far ignored is the role they may serve in diagnosis. The presuppositions and the factual content of students' explicit questions, even when false, can convey to T what information has been successfully transmitted and stored in S's background knowledge by moves (i) and (ii). This dual role of students' questions is not unlike the function of bidding in bridge.

Let us then turn to the various questions T may direct to S, move (iii). These questions, as we have indicated, subdivide into two different kinds used for two distinct purposes. We have already noted the semantical peculiarity of these questions; that is that they are not, on the surface at least, information seeking. We will discuss a real teacher's questions in more detail in Chapter VIII. Nevertheless, their functions in dialogical strategy are readily grasped and may be discussed here.

The first use of move (iii) involves what has been called diagnostic questions. These questions represent a relatively inexpensive means of determining the transmission noise in moves (i) and (ii), S's pre-existing background knowledge, or S's purposes (special interest). The most effective and certainly most economic diagnostic strategy is for T to make a random scan of S's knowledge. Move (iii) is crucial to the successful planning and implementation of teaching.From a game theoretical viewpoint it is in the best interest of the student to answer honestly provided the cooperative nature of the game remains unchanged. As long as the game remains in close proximity to its ideal cooperative state the diagnostic purpose of teachers' questions are served equally well by false (but honest) answers as by correct ones. Generally it would be best to suppress guessing since it could seriously mislead diagnosis and contribute to the selection of a less effective strategy.

The cooperative nature of the dialogical game is destroyed when, instead of for diagnostic purposes, the same questions are asked for the purposes of evaluation. The use of identically structured questions for both diagnosis and evaluation sends out contradictory signals that are difficult for S to distinguish. The same logical/semantical structure characterizes both purposes; diagnostic and evaluative tests are virtually interchangeable. (See Chapter VIII.) The use of move (iii) for evaluation changes the structure of the dialogical game, increases the expense of the move and reduces its payoff. These difficulties may be traced to the nonstandard semantical character of these questions. Perhaps this is why some cultures dispense with such questions altogether.

A second use of move (iii) is teachers' pedagogical questioning. Information ordering and eliciting questions are especially effective when used to replace or aid in making deductive moves, move (v). The former are generally less expensive than the latter. Pedagogical questions may be used to organize students' knowledge and elicit tacit knowledge. One form of tacit knowledge is the deductive consequences of already existing knowledge. No one is logically omniscient; no one knows all of the logical consequences of all they know, believe, see, remember, etc. By asking ourselves the right questions, we can learn the consequences of what we already know and believe. In order to think effectively it is also helpful to have knowledge organized in a readily accessible form. Pedagogical questions (questions that require students to organize what they know, elicit tacit knowledge and draw inferences) help the students draw their own conclusions - think for themselves. This use of move (iii) is what is known in legal jargon as asking *leading* questions, that is, questions that call for an inference on the part of the answerer. Pedagogical strategy suggests a great many possibilities for the effective use of teacher's questions.

We have so far ignored one other thing about the varieties of move (iii). Just as the content and presuppositions of students' questions sometimes signal information to the teacher, so also do teachers' questions sometimes convey information to the student. Teachers will typically ask such questions when they want to provide a "hint" to their students.

Steps of deduction, move (v), are very inexpensive in terms of material, but they can be prohibitively expensive in terms of time. This occurs when S cannot carry out the inferences necessary to reach the desired conclusion. The judicious use of moves (i) and (iv) in conjunction with (ii) can do a great deal to reduce the time cost of making this move.

The inability to draw an inference is an intellectual predicament, a gap between that previously proven and that yet to be proved. T's answers, (move (i)) to S's questions (move (iv)) can provide the missing premise; or perhaps T might ask a leading question, (move (iii)). If we care to, it is even possible to conceive individual steps of inference as answers to questions students pose to themselves. It is not necessary to go to this extreme; it is enough merely to realize that the ability to draw conclusions for oneself is not only indicative of, but partially constitutive of *understanding* what one knows.

Understanding, the ability to use knowledge, is a higher order teaching goal that has so far been ignored. By first modifying the object of the dialogical game and exploiting pedagogical strategy to the fullest we can say a great deal more not only about teaching for understanding but teaching how as well. First though we need briefly to examine moves (ii) and (vi).

These moves need not detain us long. Move (vi) is very costly. Teachers must first themselves be students. To educate them is expensive. We must build colleges, staff and maintain them, shelve libraries and spend a great deal of time teaching teachers. What justifies the expense in terms of the dialogical game is that it greatly reduces the cost of moves (i), (iii), and (iv), contributes to the reduction in the cost of (v) and could even eliminate (ii) altogether. Move (vi) may well be the foundation of the entire dialogical game.Student study moves cost time and effort. And insofar as moves (i), (ii), (iv), and (v) can serve equally well, it might be advisable to avoid making this move as much as possible. In any event, whenever this move is resorted to, it should be used in conjunction with these other moves and made to dovetail as much as possible with them.

There are other modifications involving various costs and payoffs that can alter optimal strategy that were not discussed earlier. For instance, if the costs of moves (ii) and (v) were to become very expensive the role of M would be minimized. This would convert the entire pedagogical context into exclusively teacher-student, question-answer dialogues, the premier example of which are the famous Socratic dialogues. There the burden is born almost exclusively by teachers' questions. It would be interesting to consider the effect on optimal strategy of minimizing or maximizing the cost of different moves in different combinations. For example what would happen if the costs of moves (ii) and perhaps even (v) were to become miniscule?

One modification fundamentally restructures the whole dialogical game. This modification involves changing the basic goal of the game. Rather than the acquisition of information we might decide that the purpose of the game is to prove some conclusion C. This makes move (v) the focus of the dialogical teaching game. When C involves solving some practical problem we have a model of teaching practical skill and know-how. This modification draws us close to Dewey's notion of learning as problem solving. To solve a problem is to answer a question; e.g. "What must I *do* to double a square?" or "What must I *do* to improve my back-hand?" Both questions are satisfied when the students have been shown *how* it is done, and the student can truly say, "I see that," "perceive that," etc. (Macmillan and Garrison 1986). Here as elsewhere it is probably strategically effective to break down the big question into a series of smaller questions that ultimately provide a complete answer through a series of partial answers each complete unto themselves. Indeed, all of the options of pedagogical strategy may contribute greatly to teaching various kinds of practical skills and techniques. The interchangeability of deductive and interrogative moves looms large in practical education. The most ancient and clearest example of teaching how is teaching someone what they must do to solve a problem in geometry. The most famous instance of such a lesson will provide us with an example in the next chapter.

Erotetic Teaching Strategy And A Theoretical Foundation For The Study Of Teaching

In Chapter I we issued a plea for theory in research on teaching. In the chapters that have followed we have tried to respond to this plea with an Erotetic Theory of teaching. In this chapter we have bolstered this with proof theory, game theory, and deductive heuristics. The result has been a general theory of questions and answers and question-answer dialogues powerful enough for us to take a detailed look at some of the *a priori* strategical possibilities for classroom teaching. Because of the mathematical rigor of game theory the consequences of some of the maneuvers in dialogical strategy may be calculated in advance.

It is interesting to consider what happens when certain moves become very expensive or very inexpensive. If, for example, moves (v) and perhaps (ii) become very expensive, the role of the Oracle is minimized and we get a standard question-answer dialogue. If the situation is reversed and the cost of (i) and (iii) becomes very expensive then we obtain a dialogical counterpart to those situations that do not involve a teacher *per se*. These situations range from field trips to the most refined products of educational technology. What strategies are economical, which moves become more economical and which more expensive? The forest becomes dense and dark here; these issues are not resolvable *a priori*; what is called for is the light of empirical research.These observations are intended as a sampler; we have only considered the most obvious and hence somewhat trivial instances. Still the possibilities are evident. Consider another example. What if we change the goal of teaching entirely? Instead of making the acquisition of information by S the goal, suppose skill in deducing conclusions from premises is the end of teaching. Here students are encouraged to solve problems (draw conclusions) either from what they already know and whatever they might obtain from questioning the oracle. The result is a model quite close to John Dewey's idea of learning as essentially a problem solving activity wherein "the [student's] question naturally suggests itself within some situation or personal experience" (Dewey 1916, p. 155).

One of the remarkable aspects of an erotetic game-theoretical approach to teaching is that by considering the various strategies for optimizing payoffs we are able to obtain valuable insights into the *rationale* of teaching without recourse to psychological concepts. This is not an attempt to reduce the psychology of teaching to the logical theory of erotetic games. Our purpose is not to eliminate psychology from research, but rather to eliminate pseudoproblems in psychology, i.e. logical problems disguised as psychological problems. This maneuver seeks only to place difficulties where they belong, thereby contributing to better delineating the proper domain for empirical research on teaching.

Logical investigation prepares the way for psychological investigation. Psychological research that originates in and follows from carefully conducted logical studies, profits not only from the methodological guidance and clarification it receives, but from the heuristic suggestiveness of logic. Logic does not prescribe the laws of thought, but the logical surface structure provides valuable clues to what may reside in the psychological depths. We rather doubt that game theory exhausts all there is to the concept of motivation, for example, but the logical surface has the advantage of being immediate or readily inferable; not so the psychological deep structure. One final comment. The erotetic game sets optimal strategies only theoretically. Empirical research is required to determine the best strategy in any given application. Such research is required to determine (1) the real cost of the various moves, (2) the actual payoffs, (3) the true transition probabilities. These factors vary according to time and place; but the general rules of game theory are enduring. The laws of the erotetic theory of games it appears are safe from the sort of Cronbachian (1975) decay that plagues many of the laws of psychology.

CHAPTER VII

SOCRATIC AND EROTETIC TEACHING

This chapter applies the principles and insights of erotetic theory, especially those of erotetic strategy, to the analysis of teaching. We select for this purpose a teaching episode by the most famous of philosophical teachers. From among the many dialogues in which Plato immortalized his favorite teacher we pick the dialogue between Socrates and the slave boy for closer erotetic scrutiny. We shall use the translation of W.K.C. Guthrie (1956).

The primary question of the *Meno* is posed by Meno himself, who opens the entire dialogue by asking "Is virtue something that can taught? Or does it come by practice? Or is it neither teaching nor practice that gives it to a man but natural aptitude or something else?" (*Meno*, 70a) To this Socrates replies that he does not himself know what virtue is, much less whether or not it can be taught. This problem is never settled; instead Meno and Socrates resolve merely to assume by hypothesis that "If there exists any good thing different from and not associated with, knowledge, virtue will not necessarily be any form of knowledge;" and therefore since "A man is not taught anything except knowledge" virtue cannot always be taught. Proceeding from these partially hypothetical premises, Meno and Socrates conclude that "Virtue will be acquired neither by nature nor by teaching;" but rather, "Whoever has it gets it by divine dispensation. . . " (*Meno*, 100a).

Before reaching the hypothetical stage of their discussion Socrates and Meno encounter "Meno's paradox": the Sophistical argument that it is impossible to learn or discover anything. Socrates describes the paradox as the "Trick argument that a man cannot try to discover either what he knows or what he does not know. He would not seek what he knows, for since he knows it there is no need of the inquiry, nor what he does not know, for in that case he does not even know what he is to look for" (*Meno*, 80c).

The solution to the paradox involves Plato's theory of Ideas (Forms), and with this the theory of Anamnesis (that learning is recollection). The latter theory leads Socrates to the conclusion that "Knowledge will not come from teaching but from questioning" (*Meno*, 85d). Plato's metaphysical and epistemological theories are not our concern here. We are more interested in the erotetic structure of the dialogue and the strategy Socrates employs when he demonstrates his thesis by *showing* the slave boy how to double the area of a square. What we say will speak to the philosophical issues involved, but only in a muted voice, for that is not the focus of our analysis.

In terms of the dialogical teaching game and its strategic possibilities (see Chapter VI, above), Socrates' dialogue with the slave boy is far from typical; rather it represents a very special limiting case. It is a limiting case because the theory of Anamnesis reduces the role of moves (ii) and (vi) to zero, effectively eliminating them. This has the effect of shifting the burden of the entire dialogical game onto questions and answers, moves (i), (iii) and (iv) and steps of deductive inference (v). In Socratic dialogues, questions by the teacher become especially accentuated.

This dialogue is special because it represents a distinct modification of the usual goal of the dialogical game. Rather than knowledge acquisition, the goal of the teaching episode in the *Meno* is for the student (the slave boy) to prove a certain predetermined conclusion. This brings the various layers of pedagogical strategy to play in a prominent way. The geometrical problem to be solved, stated as a question, is "How can we double the area of a square?" or equivalently "What must one *do* to double the area of a square?" Worded either way the student is expected to be able to answer the question by the conclusion of the dialogue; indeed, the answer *is* the conclusion of the dialogue. One might well doubt the authenticity of this question. (See McClellan 1976, pp. 26-28.) Is it the student's question at all, the one the boy epistemologically ought to ask? After all isn't he merely a slave of whom obedience, if not interest, may be demanded? Whose intellectual predicament is it anyway? Ideally the question should be the student's question, his intellectual predicament. One of the remarkable things about this dialogue is that Socrates does succeed in making the question the student's question, in "motivating" him.

The stage is set, beginning at 82b, by Meno's selection of a slave boy apparently at random. Socrates first asks if "He is Greek and speaks our language." The latter half of this question is asked in order to assure semantical contact between questions and answers. An affirmative answer is a necessary prerequisite for verbal or written dialogue. Socrates opens the dialogue with two questions as complex and difficult to analyze as they are easy to ask:

Q_1: Now boy, you know that a square is a figure like this? (Spoken as Socrates draws a square in the sand and *points* to it. See Figure 1.)

A_1: Yes.

Q_2: It has all these four sides equal?

A_2: Yes.

These two questions are capable of carrying out a number of erotetic functions simultaneously. It is this functional ambiguity, characteristic of the entire dialogue, that has led to so many conflicting interpretations over the centuries. In what follows we accept this ambiguity as being, in part, indicative of all works of enduring worth. Our goal is to eliminate confusion and clear the way for deeper interpretive understanding.

On one erotetic level, to start in the middle, these opening questions clearly serve a diagnostic function. By answering in the af-

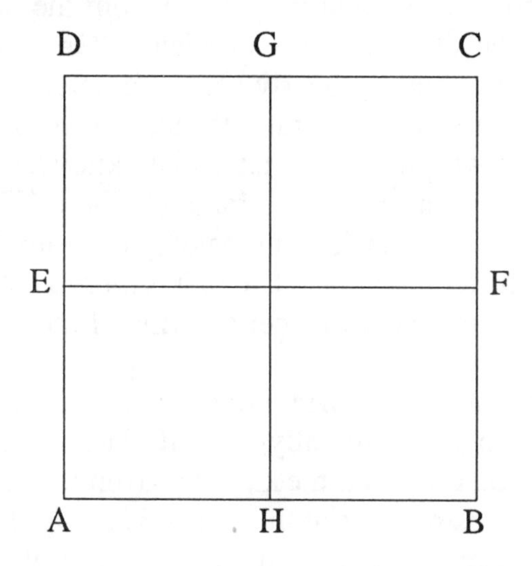

Figure 1. (*After Guthrie* [1956] p. 131)

firmative the boy establishes that he (already?) (a) knows the name of the figure, (b) has perceptual acquaintance with a square (e.g. he could pick it out of a "crowd" of geometrical figures), and (c) knows its description (or more exactly its definition). Interpreted this way these two questions enable Socrates to determine the boy's current state of explicit background knowledge and thereby his intellectual predicament regarding the problem to be solved.

Moving up one level these questions could be taken as serving an information ordering and eliciting function. Matters become stickier here. Suppose, to begin with, that the boy already knew well what a square was from a previous lesson. Perhaps his father was the family tutor. It just happened that the boy didn't have his "mind" on it at the time and couldn't *recall* the correct answer regarding what he already knew. On this occasion the questions merely helped the boy elicit tacit knowledge about facts he already knew. But suppose the boy had never seen a square, didn't know what to call it, and could under no circumstances describe it. In this case the knowledge could only come from the boy's *innate* recollection of the eternal forms or ideas. On this view the knowledge elicited is not, properly speaking, tacit but rather innate. This is of course what Socrates means to show as a result of this dialogue and why he claims later that "knowledge will not come from teaching but from questioning." On our account the two kinds of knowledge, tacit and innate, along with their means of recall are two different things. The first is more a matter of cognitive psychology, the latter of a certain kind of metaphysics and epistemology.

It is likely here at the very beginning of the dialogue that these questions function diagnostically, and if they do function as knowledge eliciting questions the knowledge elicited is tacit. The reason we say this is because at the start it is likely that Socrates merely wants to know where to begin, what the boy already knows explicitly or tacitly, and what his background knowledge is. One reason that Socrates himself probably does not see these questions as eliciting innate knowledge is that he does not pause, as he will later, to give the "technical name" of the geometrical figure. He assumes the boy already knows it. When innate knowledge of something like a square or a diagonal is elicited it is necessary to state the name men call it, for this name is different in different languages and at different times although the object itself is eternal. Since the boy, it appears, is familiar with the name it is unlikely that Socrates has drawn out any kind of innate knowledge.

In all likelihood nothing is being elicited here at all. The boy simply already knows and can recall on his own what a square is, a fact affirmed by these diagnostic questions. This does not mean that these same questions do not have an information *ordering* function. Indeed, the diagram drawn in the dust perceptually organizes what information the boy has at his disposal in a very effective way. The lesson starts with the figure ABCD and, as new knowledge is acquired, it is organized in its turn as the figure becomes increasingly better articulated and filled in. The figure serves to diagram and reflect externally the student's internal acquisition and organization of knowledge.

A third function of these questions is that they may actually be used to convey knowledge by means of their content and presuppositions. It has been suggested by many critics of this dialogue that Socrates is cheating by providing answers disguised as questions. Another way of looking at this is to see Socrates as providing the answers to the questions as he asks them. A variant on this may be that these questions are actually disguised answers to questions that the boy epistemologically ought to ask given the task of doubling the area of a square. We suspect that something like this variant may be going on; that Socrates is in fact teaching.

As the case may be, this third function, along with its variants, like the other two functions and their variants, remains a constant possibility throughout the dialogue. Sometimes, as with the diagnostic function here, there are clues as to which function is exemplified or at least is most prominent at the time. All three layers exist as erotetic possibilities in almost every question Socrates asks the boy. Moreover, these functions can and frequently do occur simultaneously; they interact with each other in exceedingly complex ways difficult to recognize, much less to analyze. We believe Socrates is actually teaching - erotetically. This does not mean that we think Plato is deliberately deceiving us (or that Socrates deceives Meno), only that they were unaware of the erotetic complexity of teaching dialogues. We do not want to prejudice the matter too much, so let us just say that in what follows we are attempting to draw as much as possible from this dialogue without committing ourselves to a Platonic metaphysics or epistemology. Stated differently, we intend to interpret this dialogue as a teaching dialogue.

In what follows, the reader must stay alert to the different interacting levels of questions and their variants. As we proceed with our analysis we will call attention to those functions that seem to us to predominate at the time along with any strategical maneuvers or other moves that deserve further comment.

So far in our interpretive analysis, we have spoken of diagnosing explicit knowledge and eliciting tacit and innate knowledge. There is a third kind of knowledge that may be elicited, call it *implicit* knowledge. It is most likely that eliciting implicit knowledge is the purpose of the next question:

> Q3: And these lines [i.e., EF and GH] which go through the middle of it are also equal?
> A3: Yes.

This question establishes a presupposition which is crucial to the final proof. As before, the same question may fulfill three different functions simultaneously. The question could be a diagnostic question. Maybe it is actually conveying knowledge, or even carrying out both of these functions at once - it's hard to tell. Certainly it is an information *ordering* question, as can be seen by the diagram. We think it is also an information *eliciting* question.

The equality of EF and GH is *implicit* in one's knowledge of a square, although it needs to be extracted in order to be made explicit. That EF and GH are equal is a logical consequence of the definition of a square. Now no one is logically omniscient; that is to say, no one knows all of the deductive consequences of everything they believe. There is no finite limit to what we can know. In any case science and everyday life confronts us with problems whose solutions could be obtained if only we properly apply what we *already* know implicitly, if we could only carry out the deductions or ask ourselves the right questions. What is required is that we activate knowledge of what we already know implicitly.

The non-trivial aspect of any geometrical proof is the introduction of auxiliary constructions. The lines EF and GH represent just such constructions. As we noted in an earlier chapter, such constructions can come to exceed the recursive power of any Turing machine to perform. The auxiliary constructions here are a great deal less than all that; nonetheless it is easy to imagine that they might never occur to the boy unless he asked himself the right questions - or someone asked them for him.

Once it has been established that the boy knows the definition of a square, the answer to the question of how to double it is already present. The actual proof, it might be added, is helped in no small way by having perceptual acquaintance with the figure, although it is not strictly speaking necessary. Knowledge of the name contributes little or nothing.

In electing to interpret this dialogue as an instance of erotetic *teaching*, it has been necessary to lay the Platonic doctrine of anamnesis and innate ideas aside. The preceding discussion now allows us to put something in their place. Let us say for the sake of the following interpretation that, rather than it being *innate* knowledge of the eternal forms that is elicited by Socratic questioning, it is the pre-existent knowledge *implicit* in the boy's understanding of squares and their properties that is called forth.

By choosing to interpret this dialogue as drawing out the deductive consequences implicit in what the boy already knows about squares, we are thrown back on the final modification introduced in chapter VI and the strategic considerations that accompany it. Recall that this modification placed tremendous weight on deductive steps by the student, move (v) of dialogical strategy, and teachers' information ordering and eliciting questions, move (iii). Students' own inferences along with teachers' "leading questions," as we have called them, loom very large in this scheme. One more thing: "How" questions are typically dealt with by detailed descriptions or by simply showing someone how it's done. Ostensive answers are important here, hence the prominent role of the geometrical figure, an empirical instantiation of the problem to be solved, as an *aid* in teaching how.

Conceived in this way, Q3 and its affirmative answer is a step, in the sense of being a partial answer, toward solving the larger problem, or question, of how to double a square and at the same time a complete answer to the question explicitly asked by Socrates. We should not lose sight of the fact that all Socrates is doing, or should be doing, is asking the questions the boy ought to ask at this stage, given the intellectual predicament of not knowing how to double the square. Presumably, the answers come from the boy's implicit knowledge of squares. We assume that Socrates only asks the questions that the boy himself epistemologically ought to ask.

The next question also establishes an important presupposition, although of a somewhat different and abstract nature.

> Q4: Such a figure could be either larger or smaller, could it not?
>
> A4: Yes.

It would have been enough to establish that the area of the square could be doubled, but all in all, the more general principle is just as easy to establish. Besides, in what follows, the boy will learn, and not by accident, how to quadruple and triple a square. Once again we are reminded of the multilayered and variant possibilities of this question and the different types of knowledge it may manifest.

The next set of questions, (Q5-Q10), are all leading questions. They call for an inference on the part of the boy. This chain of questions establishes that the area of a square double the size of the original will be eight feet. The series ends with a question about the length of the sides of a square with such an area.

Q10: Now then, try to tell me how long each of its sides will be. The present figure has a side of two feet. What will be the side of the double-sized one?

A10: It will be double, Socrates, obviously.

The boy's last answer is of course false. He has made a mistaken inference although he does not yet recognize it.

Since Q10 doubles as a diagnostic question Socrates is aware of the error and after a short consultation with Meno, he sets out to correct it. First though, Socrates confirms the boy's intellectual predicament by asking a more discerning diagnostic question.

Q11: You say that the side of double length produces the double-sided figure? Like this I mean [pointing], not long this way and short that. It must be equal on all sides like the first figure, only twice its size, that is, eight feet. Think a moment whether you still expect to get it from doubling the side.

A11: Yes, I do

This firmly establishes the boy's intellectual state of knowledge and his immediate intellectual predicament along the road to arriving at the solution of the problem. In what follows, (Q12-Q18), Socrates asks questions that allow the boy to draw closer to the final solution while at the same time slowly allowing him to *see* his own mistake.

In course of asking his questions figure 1 is augmented to obtain figure 2.

By pointing to this figure Socrates can show the boy his mistakes while at the same time organizing the information elicited by this segment of the dialogue. Although not strictly necessary, empirical instruction has been long recognized as valuable in mathematical discovery and teaching. In other circumstances such instruction is indispensable.

The dialogue concludes with the question,

> Q18: So doubling the side has given us not a double but a *fourfold* figure?
> A18: True

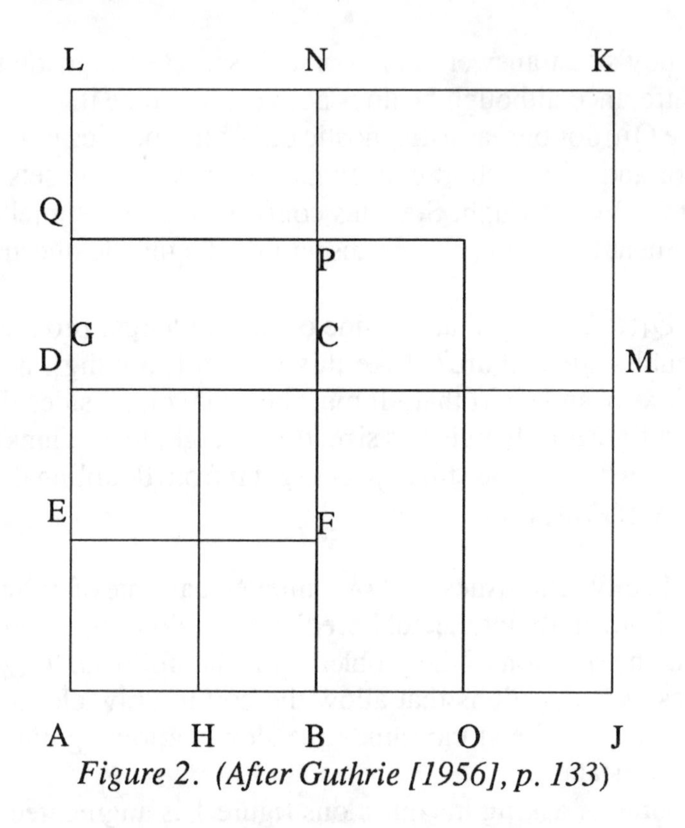

Figure 2. (After Guthrie [1956], p. 133)

Rather than allowing the subject matter alone to dictate the pedagogical order, Socrates relies also on what the boy already knows, or thinks he knows, i.e. believes, in this case wrongly. It is important to realize that teaching must accommodate the *student's* intellectual predicament including his or her mistaken beliefs. Recall that for the purpose of selecting the best erotetic teaching strategy, false but honest answers are as good as true answers and far better than a mere guess. This is why Socrates reminds the boy in the next chain of questions (at Q24) to "Always answer what you think".

The boy's affirmative answer to (Q18) indicates that he no longer believes his earlier answer, (A10), to be correct. Although the boy does not know how to double a square, he is closer than before even though he doesn't realize it. What the boy has learned are some of the deductive consequences of doubling the side of a square, one of which is that it increases the area by fourfold rather than two as he had thought.

The next interrogative chain, (Q19-Q26), draws additional inferences from what the boy has learned about the square that result from doubling the side of the original square. The focus of this bit of dialogue is on the fact that an eight-foot figure, the area sought, is double the original figure and half the area of ALKJ. This draws the boy closer to the correct solution, but once again the chain ends in an error on the part of the slave boy.

> Q25: Then the side of the eight-foot figure must be longer than two feet but shorter than four?
> A25: It must.
> Q26: Try to say how long you think it is.
> A26: Three feet.

In what follows, (Q27-Q32), Socrates again shows the boy, using the square ALKJ, that the area resulting from a three foot square is nine square feet. For a third time Socrates asks:

> Q32: Then what length will give it? Try to tell us exactly. If you don't want to count it up, just show us on the diagram.
> A32: It's no use, Socrates, I just don't know.

At this point Socrates breaks off the dialogue to register an important point regarding motivation. Socrates remarks to Meno that whereas before the boy thought he knew how to double a square now he is "perplexed" and that by "perplexing him... we have done him no harm," that "in fact we have helped him to some extent toward finding out the right answer... for now he will be quite glad to look for it" (*Meno*, 84b). The notion of perplexity here is far more logical and epistemological than psychological, having reference not to drives, need reduction and whatnot, but rather the student's (boy's) intellectual predicament. As such the motivation is entirely cognitive and internal, growing out of the subject matter. This is not to deny the existence of external and/or psychological motivation, only to point out that they are not absolutely necessary or desirable. The boy is a slave and eager to please it seems. This does not mean that he is intrinsically motivated; he might be uninterested and remain so forever. If this is the case it will not be the teacher's fault. Socrates has done just about everything that can be done to motivate an interest in the subject matter for its own sake.

Having made these observations to Meno, Socrates rubs out the previous figures and starts the lesson anew.

> Q33: Tell me boy, is not this square four feet [ABCD.] You understand
> A33: Yes
> Q34: Now we can add another equal to it like this?
> A34: Yes.

Socrates continues until he obtains figure 3 (AJGF).

The result, as the boy already knows, is a square four times as large as the original, whereas the one sought is only twice as large.

What happens next is the single most important step in solving the problem. One's interpretation of this move largely determines

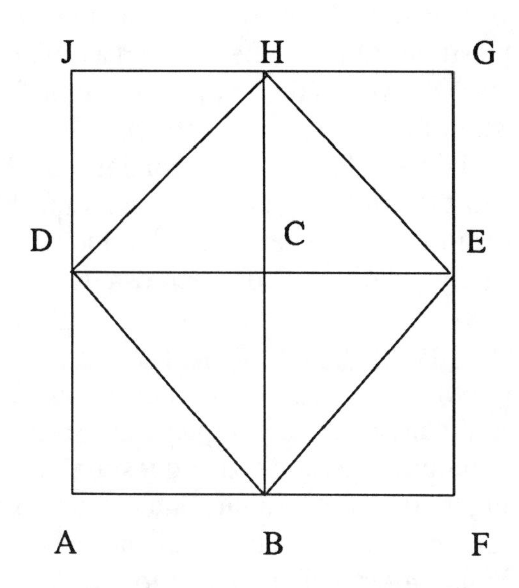

Figure 3. (After Guthrie, 1956, p.136)

one's interpretation of the entire argument. On the surface, the exchange seems innocent enough.

> Q40: Now this line going from corner to corner cuts each of these squares in half.
>
> A40: Yes.

Although this may seem trivial, it is not. The lines BE, EH, HD, DB are auxiliary constructions and, as we have said before, such constructions are the nontrivial part of any proof. In this case, once the constructions are carried out there is nothing left to be done. The square double the original is literally before the boy's eyes; all that is necessary is that he recognize it. This is the point of the remainder of the dialogue between Socrates and the slave boy.

For those who suspect Socrates of cheating, that is of disguising answers as questions, Q40 may be adduced as evidence. After all is it not Socrates who *shows* the boy *how* it is done? Isn't it his hand that carries out the nontrivial construction leaving the boy with the totally trivial task of merely recognizing what has already been done? Doesn't the old adage about actions speaking louder than words apply quite literally here? It is hard to resist this view of things, especially when we consider ostensive answers to students' questions.

Interpreting this passage along the lines suggested by Erotetic Theory seems to reinforce this perception of things. It is easy to interpret (Q40) and many of the other questions Socrates asks the boy as merely rendering explicit the question the boy epistemologically ought to ask given his intellectual predicament regarding the problem to be solved. If so, we may interpret his *actions* as providing an ostensive answer to the boy's questions by means of the auxiliary constructions. It is tempting to view the entire dialogue this way, only rather than showing the boy the answer, the answer sometimes comes in terms of the presuppositions and content of the questions. For our purposes it is enough that the dialogue can be interpreted this way, for we want to say something about teaching strategy and not Platonic philosophy.

Erotetic theory does not necessarily have to prejudice itself against a Platonic interpretation of this dialogue. We may view (Q40) as a leading question the answer to which the boy himself provides from his own background knowledge including all that he has learned so far. It is not at all difficult at this stage of the dialogue to imagine that if the boy was left with the problem of how to double a square he would soon light on the answer himself, that before long he would ask himself the right question; e.g., "what happens if I cut each square in half at a slant like this?" Again we're not sure that something like this is not happening here and throughout the dialogue.

The remainder of the exchange requires little comment. Once these final constructions are carried out it is only necessary for the boy to draw together what he already knows (in one way or another) in order to see, or recognize that the answer has been presented to him. The dialogue ends with an instructional move.

> Q50: The technical name for it is 'diagonal'; so if we use that name, it is your personal opinion that the square on the diagonal of the original square is double its area.
> A50: That is so, Socrates.

Recall the very first question of the dialogue. There, as we noted, Socrates assumed that the boy already knew the "technical name" for the geometrical figure (square). Here, since knowledge of the diagonal and the double square is supposed to be innate and not on this occasion tacitly or implicitly known, it is useful although not strictly necessary to *teach* the boy the name of the object. This act of teaching would be perfectly acceptable to Plato and Socrates since the particular name people at some time happen to call the diagonal line is not itself an eternal form (idea) and hence not a matter of recollection at all.

Socrates' comments to Meno at the conclusion of this dialogue are interesting. First of all he wins Meno's assent, if not our own, to the claim that the boy had not "answered with any opinions that were not his own," but rather "were somewhere in him" (*Meno*, 85c). We are inclined to concede this, only we would like to suggest that the opinions that were "somewhere in the boy" may have been *implicit* or *tacit* in his background knowledge rather than innate.

Socrates goes on to remark, "These opinions being newly aroused, have a dreamlike quality. But if the same questions are put to him on many occasions and in different ways, you can see that in the end he will have knowledge on the subject as accurate as anybody's" (*Meno*, 85 c-d). In this statement we have an outline for the role and structure of *drill* in erotetic teaching. The teacher needs to repeat the answers and oftentimes the questions should be explicitly asked in a number of different ways.

In this chapter we have examined the erotetic teaching strategy used in what is perhaps the most famous teaching dialogue ever recorded. We have found that the teacher selected a very special limiting case as his strategy. Nonetheless, far from being pathological, the strategy chosen seems remarkably well adapted for answering "how" questions and teaching problem solving. In fact we believe that the strategy chosen may well represent an ideal limit for all such kinds of teaching.

This strategy is easily adapted to various concrete instances of teaching, much as a physicist applies the ideal limiting case of free fall in a vacuum. In such concrete instances all of the various contextual elements must be taken into consideration. For example, the buoyancy of the fluid medium (e.g. air), friction effects, the wind blowing, etc. Similarly Socrates' strategy may require modifications to adapt it to the various contextual settings of everyday classroom teaching. The most obvious modification would be to reintroduce move (ii) of dialogical strategy. Perhaps the problem, or at least parts of it, should have been assigned as homework. Deductions by the student, as we indicated earlier, are very inexpensive in terms of materials. In this teaching episode all that is required is a stick and some dust. However, as we have also noted, such a move can be prohibitively expensive in terms of time. Teacher's questions may reduce this cost as they do in this example. Reintroducing move (ii) may increase material costs substantially; e.g. books (scrolls, tablets, whatever), laboratories, etc. Such initial increased outlays might lead to higher long term payoffs. It would also be nice if the student would be permitted to ask a few explicit questions for himself, or was at least occasionally consulted as to what he thinks ought to be done next. As matters stand, the student's questions must be inferred. Socrates, we think, has assumed these questions are present. After the motivational move (at 84a) it is likely that the boy has some questions. Shouldn't the boy be given the chance to frame them for himself?

In using the Erotetic Theory to interpret this dialogue we have tried to leave all the layers of meaning open for further inquiry and alternative viewpoints. We have chosen one way of looking at the dialogue that we find interesting and revealing. The same theory will support a number of alternative viewpoints; for example, erotetic theory permits a reading using the Platonic theory of forms and anamnesis. This does not mean erotetic theory will support *any* view. Our task has not been to distort the text, only to reveal hidden pedagogical meaning. Interpretive understanding has been our goal throughout this work; but so too has rigor and precision as well as prediction and control. We believe this erotetic analysis of a well-known teaching episode has drawn us closer to this end.

CHAPTER VIII

TEACHERS' QUESTIONS

In this chapter we shall examine two transcripts of actual classes.[1] There are two purposes for this; first, we shall examine some problems of teachers' use of questions in real life, showing how they differ from other uses of questions; second, we shall try to demonstrate that an erotetic analysis of teaching can be fruitful even in cases where the teacher is not intentionally structuring a lesson erotetically. If we are right that viewing teaching as question-answering is conceptually right and empirically fruitful, it should be so in cases where the teacher is not out to prove a theory. The examples will not definitively settle the issue, but they should support the theory's applicability to ordinary teaching.

It has long been recognized that the questions teachers put to students are not governed by the same conditions as everyday information-seeking questions. As we have said above, from such an information-seeking question, a listener is entitled to infer these conditions:

> 1. That the questioner does not know the answer to the question.
> 2. That the questioner wants to know or find out the answer.
> 3. That the questioner believes that there is an answer.

1. This chapter along with the associated transcripts is a revised version of Macmillan (1988), appearing in Dillon (1988). It is used by permission of the editor and publisher.

4. That the questioner believes that the auditor can answer the question.

In addition, each question has a set of semantical and logical presuppositions; for a question like "Who lives in that house?" we can assume that the questioner

a. believes that someone lives in that house.
b. believes that it is a house (and not a stage-set, for example).
c. believes that the form of the answer will include mention of one or more individuals who live in the house.

Teachers' questions share the latter group of presuppositions with ordinary information-seeking questions, for these are formal, in that the form of words makes no sense if they are not met. But they do not share the former group of presuppositions. In everyday information-seeking contexts, we would not be able to understand the person who asked "Who lives in that house?" knowing that it was Governor Bob Martinez (assuming here that we are driving on Adams Street in Tallahassee, Florida in 1987). In a pedagogical context, however, it is expected that the teacher knows the answer to his or her own questions.

This feature of teachers' questions makes them seem, at first glance, *inauthentic*, in that they differ so much from ordinary information-seeking questions. Some critics of schooling (McClellan 1976, for example) make the leap from this to a condemnation of any teaching that does not begin with the students' explicit questions; such condemnations go far beyond necessity. Teachers' questions should be recognized as a special case of question-and-answers, with corresponding special rules and manners. Inauthenticity in teaching is not a feature of teachers' questions in general, but rather of the multitude of specific ways that these can be misused. A parallel inauthenticity in sports is not a function of the rules of the games themselves, but of the economic and social climate in which the games are played.

When a teacher asks a question - even a simple one like "who lives in that house?" - it might be for one of at least three different purposes: (1) as a diagnosis of the students' state of knowledge or intellectual predicament, (2) as a test of the students' knowledge or attentiveness to the ongoing lesson or of their learning, or (3) as a way of carrying on the lesson. The last of these we shall call "pedagogical questions;" we will devote more space to these, for they possess unique features that have important strategic uses in teaching. The three uses may overlap in practice - and here may lie some real problems with inauthenticity in teaching.

It is important to note first that the desideratum of a teacher's diagnostic or test question often differs from that of ordinary information-seeking questions. The teacher desires to know not who lives in the house but whether the students know who lives there. The optative of such questions has the form

Bring it about that I know that you know . . . .

which differs from the ordinary

Bring it about that I know . . . .

Students generally understand that teachers' questions differ in this way; indeed in American schools at least (and perhaps even more in other cultures' schools) a teacher who asks a question for which the answer is not known is viewed with suspicion by his or her students.

When John Updike's teacher, Mark Prosser (Updike, 1959) responds to his own and a student's question "What does Macbeth's soliloquy mean?" by saying, "I don't know. I was hoping *you* would tell *me*," the students' response is amazement, or horror:

> But to [this class], ignorance in an instructor was as wrong as a hole in a roof. It was as if Mark had held forty strings pulling forty faces taut toward him and then had slashed the strings. Heads waggled, eyes dropped, voices buzzed. Some of the discipline problems, like Peter Forrester [who had asked the question], smirked signals to one another. (Updike 1959, p. 32)

Although there may be good pedagogical reasons for asking questions for which the teacher does not know the answer or pretends not to know the answer, under normal classroom circumstances, the teacher's questions are almost always viewed as having other purposes than information seeking.

The difference between diagnostic and test questions, as Hintikka (1982) points out, is a matter of the use to which the students' answers are put rather than a matter of their logical form. Diagnostic questions are used to discover what the students know or believe or can do in order to decide the next steps in teaching. "Who lives there?" asks Miss Dove, pointing at the governor's mansion; on receiving either the name ("Bob Martinez") or the office ("The governor"), she is in a position to decide how to proceed. If the office is the answer, Miss Dove can extend the chain of questions further with "Who is the governor?" or "What does he do?" Here the teacher's questions have an important role to play in planning the lessons to follow.

Test questions, on the other hand, have the purpose of determining the students' (and indirectly, the teacher's) success at their lessons. Upon such questions hang the students' futures; in effect, the teacher puts the responsibility for the answer upon the student and makes non-pedagogical consequences follow from the answer. Lack of a correct answer, in such cases, makes the student blamable.

The formal similarity of diagnostic and test questions leaves room for inauthenticity. If the students find that the teacher's apparently diagnostic questions are being used for test purposes, they are justified in losing trust in their teacher; the teacher who slips into such practices destroys credibility. It is a morally expensive slip, for if trust is undermined at this level, the whole point of the teacher-student relationship may be lost in the future.

In principle, diagnostic and test questions are relatively straightforward and clear in structure and function. The teacher seeks information about the students' knowledge and puts that information to use in the continuation of the pedagogical dialogue. More interesting and complex are a group of questions that might be bundled under the heading of "pedagogical questions."

Again, some brush must be cleared away; consider Miss Brodie's question to Rose Stanley in the following exchange:

> "I was engaged to a young man at the beginning of the War but he fell on Flanders Field," said Miss Brodie. . . . "he was poor. He came from Ayreshire, a countryman, but a hard-working and clever scholar. He said, when he asked me to marry him, 'We shall have to drink water and walk slow.' That was Hugh's country way of expressing that we would live quietly. We shall drink water and walk slow. What does the saying signify, Rose?" "That you would live quietly, Miss Brodie," said Rose Stanley who six years later had a great reputation for sex. (Spark 1961, pp. 20-21.)

Miss Brodie's question is not exactly information-seeking or diagnostic, nor exactly a test question, although it has elements of these. In another episode, she asks a similar question of Mary Macgregor, famous for being a stupid lump, and gets a wrong answer to which she responds

> "Plainly . . . you were not listening to me. If only you small girls would listen to me I would make you the creme de la creme." (Spark 1961, p. 22.)

The question is used to reinforce what she has been talking about, to make sure that her girls are attending to her, rather than to carry the lesson forward. It is a characteristic teacher's question, but one of little interest logically or pedagogically.

There are more interesting uses of questions by teachers; it is these that we shall explore by examining transcripts. We shall call these "pedagogical questions" for they are put to use in the lesson in order to further the study of the subject matter rather than (necessarily or merely) to test or diagnose the students' state of mind.

The optative for pedagogical questions is closer to the ordinary than to that of the test or diagnostic question. Here it says something like

> Bring it about that *we* know . . .

where *we* includes teachers and students; the question, put by the teacher, is based upon presuppositions that he or she thinks the students should have in their subject matter (or, generally, cognitive) repertoire. Pedagogical questions are one way that the teacher can make explicit the questions that the students ought to ask themselves.

If the erotetic conception of teaching has relevance to understanding what happens in classrooms, pedagogical questioning can be seen as a strategy of teaching that differs from mere recitation (where questions are asked to diagnose or test students' knowledge) or catechetical teaching (where speed of response seems to be the goal). In the latter, the optative practically leaves the normal question in the dust. It probably would be something like "Give the answer that you have memorized, so that I know that you can say it. . . ." The pedagogical question also differs from these in that it furthers the discussion without further inference or planning. It is part of the lesson rather than of the other tasks of teachers.[2]

Pedagogical questions can be viewed as information-ordering rather than information-seeking, since their point is to bring order to information that the teacher assumes the student have in hand or in mind. A full treatment of information-ordering questions will require attention to the relations between question, deductive moves, and other ways of distributing and ordering information in pedagogical contexts. (See Hintikka, 1982, and above, Chapter VI.)

HK's Lesson on Washington

The first transcript we shall consider is a ten-minute segment of a history class; the students are eleventh graders in a Parochial High School in Chicago. The teacher (whose name is abbreviated "HK") is a male with nine years experience. Thirty students are seated in a standard recitation-type classroom. They are largely male (60%) and white (83%).

Some commentators consider HK's teaching style to be less than adequate. It is easy to sneer at a teacher who asks a lot of questions, who carries on a class as a series of short interchanges

2. As an aside, the effects of this kind of question strategically and psychologically need to be investigated. The former is the concern here, but the latter deserves some comment. (1) It might be that this strategy misses badly when the teacher has the wrong assumptions; students might be made to feel stupid as a result ("What does *that* have to do with me?"); this also could be considered part of the strategic cost of the move in the dialogue. (2) It could well be that such moves shut off discussion or thought rather than structuring it; it does strange things to the *discussion*, but it is debatable whether it necessarily harms the students' intellectual development.

focused around specific well-formed questions to which he thinks he has clear answers. HK is such a teacher; he might almost be considered catechetical in his procedures, for he asks questions in machine-gun style, one question and answer leading to another very rapidly.

But we want to look at this episode in a different light; it is remarkably well-structured when one gets below the surface questions into the point of the lesson, which is comprehensible only if seen as an attempt to answer certain questions that HK believes the students ought to ask.[3]

HK begins the class with a general question, "Why did the Colonies win the war?" (This comes at the beginning of the class, before the transcript which we will deal with specifically.) He treats this as a question that the students ought to ask, not merely as a test or diagnostic question. He mentions that they will not have done research specifically on this question, showing that he intends them to think about what they have tried to find out in a new light.

Why should they ask this question, though? One might assume that they know the following sorts of things:

1. The colonies were sparsely populated, widespread geographically, and relatively poor.
2. The mother country had many more soldiers and was relatively rich, with a well-developed political tradition.
3. New intellectual, political, and moral ground was being plowed here.

3. These transcripts were used by an interdisciplinary team to provide different types of analysis of classroom discourse. We found it necessary to listen to the entire class to find out how it was structured, but the major consideration will be given to the ten-minute segment considered by all investigators. For the full study, see Dillon (1988).

One would expect a small, poor country with little trained leadership, a volunteer militia, and so forth, to do badly against the rich, well-trained armies of England. There is a real question to be asked here, and it is one that the students ought to ask, whether they know it or not, once they have learned about the background. The question serves as the focus for the whole class period.

Note that if all they know is that the colonies did in fact win, without the rest of the social, economic and political background, there is little reason for them to raise this question. Indeed, there might have been no raising of it without HK's intercession. He does not make the background explicit as he asks the question, and his first raising of it is met with a long pause. The student whom he addresses seems not to see the relevance of the question; he finally suggests that it is because of the importance of their cause.

After about ten minutes of discussion, we begin with a more specific question:

> T1:[4] What -- I want to go into another question about his [Washington's] military capabilities. What was it that made him militarily successful? What was it about his strategy that enabled him to be successful? Howard?
>
> B: He didn't fight straight out like the British. He fought behind brick walls, trees, and stuff like that.

Again, this is best viewed as a question put to the students as if it were the one that they ought to ask, i.e., as *their* question. It is an information ordering question, calling up the knowledge or information they have about Washington, but which has not previously been put to such pointed use.

4. Teacher's questions are given as "Tn", student's answers are given as "B" (for 'boy') or "G" (for 'girl'). An interchange will be referred to by the number of the teacher's question.

HK's next questions are even more interesting, for they bring in a counter-factual assumption:

T2: Did you read anywhere in the book where his army was destroyed?
B: No.
T3: That his army was destroyed -- Did you see that anywhere? That Washington's army was destroyed?
B: Not completely.
T4: Not completely? Tony?
B: They were outsmarted, but not destroyed.
T5: Well, is there a difference between those two statements?
B: Yeah, there is.
T6: Well, what is that difference?
B: To be outsmarted (-)[5] out of the larger army.
B: Other than that, if you're destroyed, you're destroyed, you're dead.
T7: If they're destroyed, they're dead? And what if they are outsmarted?
B: That means they could get by or move into better territory.
T8: So, what do you think Washington's success was as a military leader? -- taking into context, you know, tactics and so on. Tony?
B: He was always taking it step by step, he never wanted to be outsmarted.
T9: But you said he was, at times.
B: Right, yeah but then you asked what about his success.
T10: Well, why is it that he -- ah, Jim?

5. This mark, (-), indicates that the tape was unclear or otherwise untranscribable.

B: He was able to go back and fight harder; even after he was outsmarted, he was able to get 'em back on the rebound.

T11: Ah, because why? Chris?

G: Maybe he had to learn by the mistakes that happened, to learn by them and realize what it was that had gone wrong.

What is most striking about this interchange is how the counterfactual question works. It seems inevitable once it is made explicit; like a geometer's auxiliary construction, this negatively answered question brings a new dimension to the discussion, one that enables Washington's success to be seen as a question of certain kinds of strategies that would not have been obvious without the question. Its inevitability is in some sense logical, however, for it is unlikely that the students, left to themselves, would have raised it. Here we see how the teacher's information-ordering question focuses the discussion. We also may have a criterion of genius in teaching here - the selection of the brilliant focusing question. That's a subject for much further research!

After this question-and-answer exchange, leading to the student's conclusion (11G) that Washington had to learn from his mistakes and did so, HK raises a new question, again information-ordering:

T12: Well, did the colonies have a large army?

G: No, they -- I don't know.

This change is open to criticism in its context. The student's answer at llG had been incomplete; it was incomplete because unspecific; HK's follow-through might have asked for examples of his mistakes and how he learned from them. One might assume that the students know enough about the war to be able to think of examples; without them, they have only a very general understanding. Later, when HK does raise issues about specifics (at T27), the problem seems much more trivial, "How can we define aid?" He rejects synonyms ("help," "assistance") in the search for examples, but he doesn't follow through on these questions either.

Note that this criticism is only possible if we see what is happening in this classroom as a joint attempt to answer certain questions completely and fully; the secondary question about Washington's success as a general fits neatly into the major question about why the colonies won. But turning from the specifics in order to get on with the discussion may hinder the flow of thought here. More on another example of this later.

The discussion continues through the development of question T12, however. HK turns the attention to the nature of the army and its relations with colonies and the colonial congress.

> T13: Those of you who worked on that question about the army, and some of the problems of wartime government -- what did you find out about the army itself? Howard?
>
> B: Well, the colonial army was really outnumbered by the British. Some of the people -- the lack of interest by some of the colonies and stuff like that -- most of the colonies had to fight it by themselves, so they were outnumbered.
>
> T14: Well, ok, how did one get to be a member of the army? By way of the draft?
>
> B: No, more or less by volunteering.
>
> T15: By volunteering. Could congress, could congress tell the states to furnish more men?
>
> B: No, I don't think so.
>
> T16: What did congress have to do?
>
> [No response.]
>
> T17: Now, you're on the right track. What did congress have to do in order to get more men into the army?

B: I guess make it worth their while.

T18: Well, let's get away from pay. Let's assume that wages are not a factor at the moment. If the army is a volunteer army, as such, where did people volunteer to? [No response.]

T19: In other words, were there national offices established by congress to which a person went that wanted to volunteer?

B: You'd just go find where the army was, and join it.

T20: Not exactly. They went by way of what vehicle, do you know?

G: (-) their colonies back home.

T21: That's right, they went by way of each colony. So what we're saying, then, is that congress could not merely order a colony to furnish more men. It had to ask a colony for more men. And if people did not volunteer into that colonial army -- the Virginia army, for instance, the South Carolina army -- if people didn't then they just didn't. So, Washington commanded an all-volunteer army. And another problem with that army -- in addition to just sheer numbers, it's not professional as such, it's all volunteer, and people come and go as they want to. Now, when you have a person leading that type of a military group, you certainly are not leading a professional army, and something that he does with it is very important. It's not that large -- probably the most they ever had at one time was maybe 5,000 people. He never permits what little army they had to be destroyed. But that's not to say he never lost.

This passage begins with an apparently irrelevant question, "How did one get to be a member of the army? By way of the draft?" Actually, this is not totally irrelevant, since B13, one of the longest student responses in the lesson, talks about the individual colonies' militia being outnumbered. HK changes the subject in order to stress the relationship between the volunteer army and the necessity of good leadership by the generals. He does it by another information-ordering question, this time stating a presupposition that the students might believe: that the colonial army was staffed by draftees as a modern army is. The question has to be refused, and it is at B14. Then a new presupposition takes its place: that men would volunteer for the national army as they might now volunteer for the U. S. Marines. HK raises this question, again as a series of information-ordering questions about congress's role in the selection and reception of volunteers. He weasels an answers from the students finally at G20: "[They volunteered in?] their colonies back home." This sets the stage for HK to summarize what has gone on so far: the American soldiers were volunteers in each colony, giving peculiar problems of leadership and strategy that (as the students had seen in earlier parts of the discussion) Washington had solved.

At this point a student asks a "real" information-seeking question:

B: When did the draft come about? Or when was it, you know, where it was not a come and go situation?

T22: It would -- Tony, it's strictly a guess on my part, but I think that the Civil Was was the first time that there was a wartime draft.

B: Because at this time leaders, like Washington, they couldn't give orders and be mean, 'cause it was a come and go thing, and they'd just leave, so as you (-) and come out, you get more strict generals.

T23: That's right. Do you know why people left, by the way? Why, we say, they come and go? Do you know why they left?

B: To do the harvest.

T24: They went home and farmed. When it was time to harvest the crops--stop fighting, go home and take care of the home front, and when you come back, you fight again. That was even true to some extent in the Civil War; that was not terribly uncommon.

HK picks up a sub-question about the soldiers' coming and going rather than following through on the draft question or pushing deeper into the nature of leadership problems in volunteer armies. The student's question goes beyond the immediate problem of the revolutionary war, but it shows that HK's earlier moves are to the point: the students see the point about the draft, and they see what a difference it makes in the ways that war can be waged.

This passage is particularly important because it shows how the teacher can use questions not for diagnosis (although there is some of this in the interchange: "What do these students know about this? Am I right in my assumptions?"), but in order to arrange the information they have about the subject under discussion. One can criticize particular moves made by the teacher only if they are seen for what they are. The weaknesses are missed opportunities to make the material more cogent, to follow through in order to make sure that complete answers are presented. This logical type of critique should probably precede any other criticisms having to do with the nature of learning, psychological processes in the classroom, or whatever.

But HK returns to the question that was the focus of the class from the beginning:

T24: [continued] OK, so we've kind of covered leadership and some of the things that Washington brought with it. Why else did they win? Leadership is important, that's one.

B: France gave 'em help.

T25: OK, so France giving aid is an example of what? France is an example of it, obviously.
B: Aid from allies.
T26: Aid from allies, very good. Were there any other allies who gave aid to us?
B: Spain.
T27: Spain. Now, when you said aid, can you define that? How can we define aid? Greg?
B: Help.
T28: Define "help." Spell it out for me.
B: Assistance.
T29: Spell it out for me. Joe?
B: They taught the men how to fight the right way.
T30: Who taught?
B: The allies.
T31: Where? When?
G: In the battlefield.
T32: In the battlefield?
B: The allies would help if they would gain something after their help, a reward or something.
T33: Well, Greg has said that we received aid from France. I've merely indicated, what do you mean by aid? What do you mean by help? What do you mean by assistance: Those are general words. Can you spell it out concretely, exactly what that aid was?
B: Food and ammunition and stuff like that.
T34: Food, ammunition.
B: Clothes.
T35: Was there anything more important than all of those?
B: Men.
T36: Did they send large numbers of men here? Miss Edwards?
G: Money.

T37: Money. France loaned us two million dollars.
G: [Is that all?]
T37: [Continued] That was a lot in those days. We also received some from Spain. We also received some from Holland, eventually. OK so we have leadership of Washington, we have foreign aid -- now are there any other reasons why we won?

Again, we see that HK is ordering the information the students have gained through their reading, putting it into a new context. If there is a fault here, it is that he is trying to have them guess what he has in mind - money. But his questions and acceptance of the answers (apart from a confusion of definition and exemplification of 'aid') moves the discussion along, puts their knowledge into a new order.

A student then asks a question from the real world; HK, a bit startled, gives partial answers, putting off till later the discussion of the issue:

B: I wanted to go back -- after this is over, we have to repay them, right?
T38: We--yes, we do.
B: So that might mean that taxes start higher.
T39: No, oddly enough, no.
B: Who'd pay for it--Congress itself?
T40: Congress would have to pay itself in the form of gold or silver, it could pay that debt off. Now we could have reneged on that debt, but that's another chapter. We'll get back to that, in fact -- we do give that money back.

This move opens the way to new questions when the discussion moves to the first years of the American Republic. HK is up on the question, and he knows where it is going to take the discussion if it is answered now, so he puts it off with a promise. Here a question that ought to be asked by the students is asked and will provide a focus for a future discussion.

T40: [Continued] Are there any other reasons why we won? Fred?
B: The French, I think they brought some military genius - Van du Peau or something like that.
T41: Well, is that not part of foreign aid? Remember, we asked about aid? Those Frenchmen who enlisted or came here on their own to fight, would I think be a form of aid whether it was individual or not.
B: Propaganda.
T42: Propaganda.
B: Get people thinking the right way.
T43: That's right. Is propaganda important in war?
B: Control the negative ideas (-)
T: Chris?
G: I'd say yes, for getting men. (-) their own way of saying, we'll fight for you.
T44: What else?
G: Yes, (-) organized army or something (-)
T45: From the colonies?
G: Yes.
T46: Well, you said something a while ago about tactics. Could that have played a role in our winning? Could that play a role, colonial tactics?
G: They weren't predictable. [T: What?] They weren't predictable.
T47: What do you mean by "not predictable"?
G: They couldn't be -- they couldn't guess at what they're gonna do before they did it.
B: Would it be because they attacked at night?
T48: Did they?
B: Sometimes they did.

T49: Was that bad?
B: No.
T50: No - I don't mean "bad", was that different?
B: Yes.
T51: How did Europeans fight? Joe?
B: Straight out in lines, in the fields.
T52: Open in the fields? In lines?
B: In lines, straight.
T53: How did the colonials fight?
B: Behind trees.
T54: Behind trees?
B: On farms wherever they could hide.
T55: Was that fighting by the rules?
B: No. There was no rules.
T56: Is there such a thing as fighting by the rules?
B: That's what war is, though.
T57: Is war fought by rules?
B: Heck no. It's supposed to be.
T58: Golombiesky, is it fought by rules?
B: It's supposed to be. It usually isn't, though.
T59: Can you give me an example of something that would be unfair in time of war?
B: Fire, and shooting on an ambulance.
T60: What's unfair in time of war, Howard?
B: Killing somebody that's surrendering.
T61: All right. Can you name me anything else?
G: A paratrooper is falling in flames and you're not supposed to - you shoot him - No!
T62: What can't you do during time of war? Tony?
B: Kill prisoners of war.
T63: Kill Prisoners of war. Can you torture 'em?
B & G: Yes/no.
B: You're not supposed to, but they have been known to do it on occasion. [Laughter.]

T64: Is there anything else that people would look down upon a nation for, if a nation did it?
[Here ends the transcription.]

By the end of this discussion, as we can see, the class has changed topics to the nature of rules in warfare; there are rules, we find; the Americans did not follow some of them, but there are others (perhaps) that they did follow. There is another missed opportunity here, for HK does not ask what rules Washington *did* follow. The concern moves from strategic rules to the moral rules of war, and this distinction is not caught up adequately. Washington broke the rules of fighting, but one can make a case for his carefully following the moral rules, specifically those having to do with the treatment of civilian enemies and prisoners of war.[6]

The discussion does not come back to this, perhaps, because HK thinks that Washington's leadership has been covered and that this is a feature of that aspect. There's a danger in covering ground and summarizing too soon. In shutting off apparent repetitions (as he does with earlier questions about leadership at T24 and foreign aid at T41), HK misses opportunities that could have deepened the students' knowledge of that period and their own understanding of the present situation in the world.

Is there anything general to be said about this discussion erotetically? We have here a surface manifestation of a question-asking teacher; to understand the logic of the lesson we have to look below the questions to see how they serve a pedagogical point. The questions HK puts are not for *his* information seeking purposes, but rather for the students' understanding of the informa-

6. Note that these criticisms cannot be raised without some knowledge of the subject matter. In order to understand what was happening in this class, we had to review our own knowledge of the Revolutionary war! A purely formal analysis of classroom interactions, we fear, will usually miss the most important points.

tion they already possess. Ranging from a few diagnostic questions, through some general questions that they ought to ask, back and forth through specifics and generalities, HK finds ways of putting their present information into new contexts, giving it new importance for them. He asks *their* questions, and insofar as he hits them right, i.e., insofar as these really are their questions, HK moves them from a state of rather diffuse knowledge about the Revolutionary War into a state of much more highly structured knowledge; what they know is shown to be relevant to different types of issue. He also moves them back and forth from their knowledge - also diffuse - of modern war to the situation faced by the colonies. The moral points of the later part of the discussion are not to be sneezed at in this context, for they make the humanity of the situation immediately relevant to their present experience of political and military involvement.

HK is no Socrates, but he is doing what Socrates does in the slave-boy episode: by questioning the students, he helps them realize what they already know. Not a bad thing to do!

No Smoking: MK's Christian Life Class

The second transcript we will consider has a different style and content. Again, the class takes place in a Parochial High School in Chicago (a different one). The class is called "Christian Relationships;" there are twenty-eight students (43% male, all white), sitting in a standard recitation-type classroom. The teacher ("MK") is a male with four years teaching experience; the students call him by his first name ("Larry"). For reasons of space, we will not provide the entire ten-minute segment here, but rather give highlights.[7]

As before, it is necessary to go to the beginning of the class to see the point of the discussion; the discussion is harder to follow

7. Again, the source for this is Dillon (1988). The entire transcript is given there.

in this excerpt, in large part because MK did not set out to have the class discuss the issue at such length.

At the beginning of the lesson, MK notes that the topic for the day was to have been slightly different from what evolves:

> Before we go on talking about the concept of being at home, feeling welcome, feeling like you belong somewhere, . . .

he wants to consider an incident that occurred the day before. He asks a student to explain to the others. Only a garbled version comes across, but it seems that a teacher went to a place outside the school where only seniors were permitted to gather; the teacher smelled marijuana; this presumably was reported to the principal; the seniors' privileges were rescinded, whereupon the seniors threatened to walk out of school.

At this point our segment begins. MK asks

> T1: Does anybody have any strong feelings about that--about what happened--about the seniors having their privileges taken away?

This question leads to a fast-paced, garbled set of responses by the students:

> Sa:[8] I think (-)
> Sb: Seniors should get to do what they want.
> [Teacher: OK, John]
> Sc: If the seniors can't handle it, the juniors will handle it.
> Sd: Ah! (-) why can't you smoke?

8. Because of the speed of the interactions, we shall refer to student responses by letter: "Sa" is the first response, "Sb" the second, and so forth.

Se: Why can't you just smoke cigarettes at this school?
Sf: 'Cause it's illegal.
Sg: Why?
Sh: 'Cause you're a minor.
Si: Cause like at Loyola, all right, and Loyola is just as strict as this school, and seniors can smoke there.
Sj: Yeah, but they probably got a whole lounge.
[Teacher: OK]
Sk: I don't see why just because of what a few kids did, the rest of us have to suffer - it's not fair.
Sl: (-) this neighborhood.
Sm: What if you're 18?

Let us discuss this passage. We must ask what the pedagogical point of the teacher's question [T1] amounts to. There doesn't seem to be any reason for MK to believe that *no one* in this class has strong feelings about the incident. An information-seeking question of this form would admit of such a possibility. The background would be something like "Either someone has strong feelings or no one does." But MK knows that they do have strong feelings about it. His question starts the discussion onto a manageable track. It is an information-ordering question: "What are your feelings about this?" Note, however, that it is a different kind of information-ordering from HK's. The question is not "What do we know?" but "What do our feelings amount to?"

This is a doubly unfocused question: first in that it is directed at anybody rather than at any specific student; second, in that it does not limit the answer to a *particular* feeling. A focused question on this might be something like, "Barb, do you feel angry about this?" Such a question would narrow the range of response to a single yes-or-no. MK has more fish to fry, though.

The question succeeds in starting something: thirteen rapid-fire responses, garbled and heated talk among the students, not *report-*

ing feelings ("I hate it," or "I love it") but *expressing* feelings along with attempts at clarification, questions about rules and rights, and plain frustration at their lack of power in this area.

The discussion continues for considerable time along the same lines - students attempt to justify the teachers' actions in rescinding the seniors' privileges, express outrage, and so forth. At last, MK enters with a change of topic:

T14: What if somebody came by to you and said, "Listen, I'll tell you what, it's about time that you started to act Christian about this" How would you define a Christian behavior in something like this? Can it be defined?

Sa: Yeah, it could, but you'd have to think about it.

T15: OK, let's look at it in terms of the smokers. Is there such a thing as a Christian behavior in terms of the smokers? How would the smokers respond? [S: Christian behavior?] If you identify - I'm not saying that it can be, but I'm asking a question.

Sa: Let the smokers go in the john and do it, so if they get caught, it's their responsibility - it's not everybody else'll get in trouble.

Sb: That's right. We'll take our own - if we get caught, it's our fault, not anyone else's. [Teacher: OK, Barb.]

Sc: Yeah, do not smoke out there where everybody else in the school (-). You got have respect for that, you know.

T16: What's the Christian response to the people who are out there? Or who are not smokers, who are not smokers.

Sa: Tell 'em, "Man, get out of here with that." I mean, "You're gonna get us all in trouble."

T17: What's the Christian response - if you go up to somebody who's sitting down there smoking, and you say, "Hey, stop smoking," and they say, "Bug off"?

Sa: (-) wrath of God. [Teacher: The wrath of God? All right, let me - Barb?]

Sb: I was just wondering, like - I dunno, this might be kinda dumb and stuff - but, OK, like, people go up in front of the church - well, I think it's two people - OK, they go up in front of church and they always say how much Christian they are, and stuff like that. And then like, just now you were talking about - the way it sounded to me was - like you said, "What is the Christian attitude to people who smoke?" So, does that mean if you smoke, you're not a Christian?

T18: Well, I - that's a good question.

We see here that MK's question expresses the question that he believes they ought to be asking in the context of a class on Christian life, where the topic is what it is to be a member of a community. We also see how difficult it is for the students to come to any sort of conclusion about this. It is, after all, not something written, a rule to follow, but a situation to be considered in light of the general principles of Christian life. This is seen in the final student response of the transcript (after T19):

Se: I remember something you said, Larry [= MK], a while ago - I think it was right in the beginning of class [i.e., at the beginning of the term] where you were explaining Christian relations, OK, to the class. [Teacher: Yeah.] And you said that you were hitching and you got a ride or something like that. And you were sitting in the back seat of the car. [Teacher: Yeah.] And you brought out a cigarette, and the lady said, "You're defiling the Holy

Temple of God," and she offered you a stick of gum, and you said, "No, it'll rot my teeth." Same thing!

T20: That's true. . . .

It is obvious that no closure has been reached in this discussion, no question firmly and completely answered. But the discussion itself hangs on the questions that the teacher asks for the students, the questions that he thinks they should be answering. Insofar as there is a structure to this class, it is the structure given by the questions.

Interestingly, there is a final disclosure of the recurring theme, perhaps the ultimate question for such a class. After the transcript, as the class ends, this interchange occurs:

T: Is it possible that you could win by losing?

S: (Quietly) What do you mean?

T: I won't say any more about that. We'll talk about that a little later.

Perhaps MK had a hidden question or a message to get across here. The answer to the general question is to be found in "winning by losing." He puts the seeds of this idea forth as the ending - but not *concluding* - point of the lesson with a promise to return to it. They have, perhaps, reached a point in their discussion of the practicalities of Christian ethics where the question he thinks they ought to be asking is this more or less strategic point: how can one *win* and still remain true to Christian principles?

MK's lesson on smoking in the "back" comes back to the central questions of the class. It is one of those serendipitous episodes that teachers all too often miss. He turns this into an opportunity for making the central questions of Christian life come alive in a living example for his students. The logic is not as clear as in a more structured class; perhaps this is as it should be. For the questions of modern life and the place of Christian ethics in modern life are not as clearly approachable as issues of history, for example, or even more structured subjects like science and math.

Concluding Remarks

This tour through these two classes should show some of the force of the erotetic approach to teaching as an analytical and critical tool. There are several useful and promising things in the approach.

First, we can see the *structure* of classroom discourse as given in the relations among the questions students ought to be asking, the answers the teacher provides or authorizes, and the continuation of the discussion of the subject under consideration.

Second, these features can be seen in the lessons even though the teachers themselves were not consciously aware of the logical structure viewed this way. That is, these teachers can be fruitfully described "erotetically" even though they did not plan it that way. This suggests that the claim that the logic of teaching is the logic of questions is at least thus far defensible. Other analyses of teaching episodes that we have analyzed suggest the same, even where there is more "noise," where subsidiary activities take up more time in the classroom.

It does not appear that putting this theory onto the classroom descriptively does an injustice to the facts. But much more needs to be done before we are willing to assert this unqualifiedly.

Third, there is promise for more research using this model. Some of this research will be empirical, addressing the issue of how best to uncover the intellectual predicaments that underlie those questions that ought to be asked. Some will be logical analyses of subject matter with an eye to the question-answering power of different forms of presentation. Others will be investigations into strategies and tactics, combining empirical and logical elements in ways that may be specific to particular subject matters and students.

What seems most promising about this approach is the almost direct translation into practice; for the erotetic approach uses a literal, intentional conception of teaching (not a metaphor), so that the findings of research are cast in the same language as that used by teachers themselves. If we are right about this, such an approach goes a long way toward bridging the gap between research and practice. The training and evaluation of teachers might take a new turn as well; and finally, teachers themselves might find new ways of answering all those questions. Would HK or MK have been able to focus their lessons more adequately had they thought of teaching this way? We'll never know - but perhaps that kind of question can at least be raised now.

CHAPTER IX

EROTETICS, COGNITIVE PSYCHOLOGY AND THE PROCESS OF PROBLEM SOLVING

Thus far, we have restricted ourselves to the erotetic approach to teaching. This chapter will reconsider the basic ideas of erotetic logic and relate them to some of the principles central to cognitive psychology and information processing models of problem solving. The introduction of erotetic logic into the discussion of problem solving not only provides an alternative way of studying problem solving activity, but more importantly, it also provides a "logical home" for a number of the central concepts and theories found in this burgeoning area of study.

It will become evident that our application of erotetic logic to problem solving inquiry is a natural extension of the earlier discussion of erotetic teaching strategies in Chapter VI and of Socrates' solution of the geometrical problem in Chapter VII. Ours is a logical theory of *teaching* - yet we feel that at some point our theory must begin to link up with adjoining areas of educational inquiry. By displaying an intersection between erotetic logic, cognitive psychology and the study of problem solving, we may begin to see how such a link might be forged.

Problem solving has two aspects: (a) the *process*, or set of activities that guide the search for a solution, and (b) the *product*, or the actual solution. Most empirical studies of problem solving have been experimental in design and have used scores on objective measures of achievement (i.e., the product of problem solving) as the dependent variable, but more and more attention is being given to the processes of problem solving (Frederiksen 1984). A process-oriented theoretical structure for discussion of problem solving is one result of this activity.

If there is a fault of the resultant theoretical structure, it is that its developers have not attended clearly to the difference between *a priori* hypothesis and theory development and the empirical results of studying actual problem solving. In other words, the leading theorists in the fields of cognitive psychology and information processing have not attended closely to the logical structure (and background) of their theories as differentiated from the data used to support or falsify those theories. We believe that closer attention to the epistemological status of these theories could lead to cleaner studies and more useful results.

Hence, we shall examine the relevance of erotetic logic to the theoretical background of recent studies of problem solving. Our goal is not to eliminate empirical psychology from the study of problem solving, but rather to reduce the amount of work required of empirical psychology. By putting the mystery in the right place we can better circumscribe the proper limits of empirical psychology while enlisting logic as a guide to empirical research. The truly psychological can never be reduced to the logical.

Mere logic will not do as the conceptual basis of theories designed to describe and study the real world. The logical world is clean, neat and well ordered. A high level of abstraction permits precision, but often the price of precision is (empirical) content. The everyday empirical world, the real world, is commonly dirty, messy, and disorganized. But if we must be cast out of (logical) paradise, let us at least know what it is that we are missing.

Standard sentential logics are static; they are designed to display the relations among sentences (or, more technically, propositions) quite apart from how changes in the epistemological or semantic background of those sentences affect their relations. As a result, they have limited use in describing the real-world dynamics of such areas of human endeavor as learning, teaching and problem solving. Erotetic logic, on the other hand, is a dynamic logic; it is directly concerned with describing how such changes take place, with the codification of rules for assessing such situations. It is because of the dynamic nature of erotetic logic that we undertake this exploration of problem solving and cognitive psychology.

Problems, Questions and Information Theory

It is easy - even natural, perhaps - to conceive of a problem as a question that we cannot answer given the knowledge immediately available to us. Likewise it is reasonable to conceive of a solution to a problem as a completely satisfactory answer to an underlying question. Information theory, interestingly (and surprisingly?), can be used to provide a rigorous and precise formulation of this intuitive equivalence between problems and questions.

Let us begin by distinguishing between questions posed *to* some agent, called the Inquirer, by some source of problems, and questions posed *by* the Inquirer to some source of information. We will refer to the "source" of both problems and information as the "Oracle." In practice the role of the Oracle may be played by the subject matter (or curriculum), a teacher, parent, "nature", a computer, *et al.* A question posed by the Oracle to the Inquirer that the latter cannot answer constitutes a problem for the inquirer. Understood thus, a problem is any uncertain and indeterminate situation, where, in Dewey's words, whatever "perplexes and challenges the mind so that it makes belief at all uncertain . . . is a genuine problem, or question. . . " (Dewey 1933, p. 13).

In information theory the Inquirer's uncertainty is defined and measured by the number of choices or alternatives to the actual situation that are compatible with what the Inquirer already knows or believes about the world. We may, following Hintikka (1976), refer to these compatible alternatives as "epistemic alternatives."

Hintikka identifies the Inquirer's epistemic alternatives with "knowledge worlds" or "possible worlds." The number of epistemic alternatives compatible with what the Inquirer already knows (background knowledge) provides a measure of the Inquirer's uncertainty. The number of choices or alternatives to the actual situation that a given answer to a question eliminates provides an accurate measure of the answer's informativeness. So conceived, a partial answer eliminates some but not all alternatives.[1]

Using Hintikka semantics it is natural to consider questions posed by the Inquirer to the Oracle as intended to solve some problem, i.e., to answer the original question by eliminating the various epistemic alternatives to the actual situation. The Inquirer's questions to the Oracle are information seeking questions that have as their goal the elimination of uncertainty. Likewise, the Oracle's veridical answers, when forthcoming, provide information - that is, they eliminate the Inquirer's uncertainty with regard to the alternatives open to him or her.

Some general observations may be made about the Inquirer's epistemic alternatives. First, not all alternatives need initially appear equally plausible to the Inquirer. Second, it is unlikely that the Inquirer will be explicitly aware of all the epistemic alternatives compatible with what he or she knows. Rendering tacit alternatives explicit is often crucial to problem solving. Third, false belief or inadequate knowledge on the part of the Inquirer may ex-

1. Hintikka's possible worlds interpretation for questions involves us in a controversial semantics that we were careful to avoid earlier. We take up this semantics now because it allows us more readily to recognize the information-processing properties implicit in the logic of questions and answers. Our interpretation could just as well be provided with an equally satisfactory, if somewhat less straightforward, information processing interpretation.

clude the actual situation from the Inquirer's set of alternatives. In such situations the Inquirer must multiply alternatives before reducing them. This calls for productive thinking. In one way or another the Inquirer needs to learn more, "unlearn" false beliefs, or even "relearn" things that have been forgotten. Or, equivalently, gather information, eliminate noise or recover lost or poorly stored information.

The Erotetic Logic of Problem Solving: A Precedent and a Paradox

Question and answer dialogues have been considered a method of inquiry and knowledge acquisition (i.e., information gathering) since antiquity. In Plato's *Republic*, for instance, dialectic is conceived as a method of discovery, the highest form of inquiry (*Republic*, 511b). Likewise, Aristotle also takes dialectic as a means of discovery. These precedents are of more than mere historical interest, however. They provide us with one of the most immutable and intractable principles of problem solving inquiry - Meno's paradox. (See above, Chapter VII.)

Aristotle's solution to Meno's paradox consists in declaring that "There is nothing to prevent a man in one sense knowing what he is learning, in another not knowing" (*Posterior Analytics*, 71b5). The payoff of this is found in his remark, "All instruction given or received by way of argument [inquiry] proceeds from pre-existent knowledge." This idea is well ensconced in the field of cognitive and educational psychology. (See, for example, Ausubel 1968.) Inquiry is impossible for an Inquirer who is without sufficient prior information to recognize and represent the problem well enough to anticipate, however inaccurately, the eventual solution.[2]

2. It is this simple fact that makes atheoretical inquiry an impossibility (Garrison 1987).

Aristotle recognizes two kinds of pre-existent knowledge: The meaning of the word and the existence of the thing or the matter of fact (*Posterior Analytics*, 71a11-13). For example, in geometrical inquiry it may be enough simply to know already the meaning of the word 'triangle' as "a three-sided figure enclosing a space the sum of whose interior angles equals 180 degrees." On a different occasion it may be enough to know the fact that there is "this particular figure inscribed in the semicircle."

Aristotle goes on to describe one special kind of solution: "Recognition of a truth [a solution] may in some cases contain as factors both previous knowledge and also knowledge acquired simultaneously with that recognition" (*Posterior Analytics*, 71a16-18). He gives an example: "The student knew beforehand that the angles of every triangle are equal to two right angles; but it was only at the actual moment at which he was being led on to recognize this as true in the instance before him that he came to know 'this figure inscribed in the semicircle' to be a triangle" (*Posterior Analytics*, 71a,19-22). Here the solution turns on prior knowledge of a meaning and the recognition that some particular thing is an instance of it. Modern cognitive psychology takes a similar view of "pre-existent knowledge." We will return to this shortly.

As in Chapter VII, we will not concern ourselves with the deeper metaphysical and epistemological issues that arise from considering the implications of Meno's paradox. We will rather content ourselves with the fact, as well known to cognitive psychologists as to philosophers, that some form of prior knowledge is required in order to initiate problem solving inquiry. We remain agnostic as to the origin of this prior knowledge. It suffices for our purposes simply to recognize that in the logic of questions and answers "prior knowledge" appears in the guise of those established presuppositions necessary for the Inquirer to formulate information-seeking questions. Pre-existing presuppositions (a redundancy) initiate, guide and constrain problem solving inquiry by determining how questions are formulated and, on the part of the Inquirer, which questions should be asked in what order.

The prior knowledge necessary for overcoming Meno's paradox is the kind of "extra-logical" epistemological criteria, or *pragmatics*, that we discussed earlier in terms of complete answerhood; "background knowledge," we called it in Chapter IV. The connection between answerhood and the paradox should be clear. Complete answerhood is the same as the completion of the inquiry, i.e., it answers the "big" question, thereby solving the problem by hooking up what the Inquirer knows at the beginning with what he or she seeks to know.

Erotetic Logic, Cognitive Processing and Information Theory

Let us begin our investigation of the erotetic logic of problem solving inquiry by examining how inquiry is initiated.

In cognitive psychology and information-processing models, information is understood to be stored in long-term memory (LTM). Information is usually taken to be stored in the form of *nodes*. A node represents a unit of information that may be interrelated in complex ways with other nodes (Kintsch 1972; Schneider and Shiffrin 1977). Some nodes contain sensory-perceptual knowledge of facts and others store semantic or propositional knowledge. [The similarity between Aristotle's two kinds of pre-existent knowledge and the kinds of information stored in LTM is striking.] This information may be stored in LTM by highly organized and interconnected conceptual networks (Anderson 1981; Puff 1979) wherein concepts may be represented as nodes and lines connecting the nodes stand for (meaningful) associations between concepts (Frederiksen, 1984, p. 364). Since, in erotetic logic, the presuppositions of questions include their conceptual presuppositions it is not too much of a stretch to consider the conceptual network stored in LTM as comprised of the logical presuppositions of questions and their connections. Information in the form of presuppositions stored in LTM provides the requisite prior knowledge for initiating inquiry.

In an earlier chapter we spoke of the "epistemological state" of the questioner. Much of what follows turns on equating this notion with the questioner's "cognitive state," the term of choice in cognitive psychology.

In cognitive psychology, matters dealing with the initiation of inquiry are discussed in terms of the "problem representation." Newell and Simon (1972) divide problem representation into the concepts of "task environment" and "problem space." The task environment is comprised of facts, concepts and their relations that make up the problem. In our terms, this is equivalent to the question posed by the Oracle. The problem space is defined by Newell and Simon as the problem solver's (Inquirer's) mental representation of the task environment. The facts, concepts (meanings), and their relations comprising the Inquirer's problem space must initially be drawn from LTM. These constraints on the initial construction of the problem space are the same as the constraints of pre-existent knowledge described by Aristotle. Erotetics renders the notion of problem space precise, at the price of mental content, perhaps, by depsychologizing the notion of problem space, putting it rather in terms of "logical space." (Or "logical gap.")

Inquiry is initiated from the presuppositions and their relations, along with any facts available to the Inquirer. As recent philosophers of science have emphasized (e.g., Hanson 1958), presuppositions and facts are not independent; there are no theory-independent facts; rather, facts are dependent upon the theory, or more exactly on the theory's concepts that allow us to recognize and interpret phenomena *as* something.[3] Rather than speak of the

3. 'Theory' must be construed broadly here, to include the everyday assumptions built into ordinary language. Seeing an object as a tree carries with it a "theory" that includes distinctions between trees and shrubs, as well as other distinctions that exclude stones, houses and people from the category.

theory- or concept-ladenness of observation, we could (and will) follow Hintikka and Hintikka (1982, p. 66) and speak of the "question-ladenness" of observation. This notion is equivalent to the psychologist's "selective attention" but focuses more directly on the place of the attention in perception and inquiry.

In any event, the Inquirer must initiate inquiry from those presuppositions (and the perceptions that the presuppositions make possible) that constitute the Inquirer's problem space. The presuppositions in the problem space comprise the pre-existing knowledge necessary to overcome Meno's paradox and initiate inquiry.

An inaccurate, inexact, or ambiguous problem representation may make it more difficult if not impossible to solve a given problem. Consider some of the more obvious (logical) possibilities. If the Inquirer finds the question posed by the Oracle conceptually confused and/or the putative "facts" referred to by the question false, the Inquirer is likely to reject the question rather than attempt to answer it. The inquirer will dissolve rather than solve such a problem. The question, "How high can a unicorn jump?" is, in Aristotle's terms, meaningful - i.e., we know the meaning of the words. But if we don't believe unicorns exist, we reject the question. Likewise, we might find the question, "How many slithy toves did gyre and gimble in the wabe?" grammatical but meaningless; since we don't know the meaning of the words, we are skeptical, without good reason perhaps, about the existence of their referents. Do "slithy toves" exist? If we believe not, we reject the question. Similarly, conceptual confusion and false presuppositions on the part of the Inquirer can confound inquiry. Perhaps unicorns do exist and are in fact the cause of quantum leaps in physics. Perhaps an occupant of the looking glass world, say Humpty Dumpty, would know what a "tove" is. The nineteenth century logician, C. L. Dodgson explored the difficulties and dangers of inquiry in regions where our everyday presuppositions fail to hold - in *Alice in Wonderland* and *Through the*

Looking Glass. (For a philosophical discussion of these works, see Heath 1974.)

A Problem-Solving Definition of "Intellectual Predicaments"

The gap within the problem representation that lies between the Inquirer's problem space and the task environment set by the Oracle's question may be said to constitute a "cognitive gap," in the terms of cognitive psychology. The Inquirer's task is to close this gap completely. In erotetic terms, this task constitutes the Inquirer's "intellectual predicament."[4]

It does not stretch credibility too far to identify the "problem space" with the premises (in a loose sense of the term) available to the Inquirer in carrying out a deduction or proof of some theorem or conclusion. Nor is it unnatural to equate the "task environment" with the final conclusion of some logical deduction or proof. (This is, of course, what we did in Chapter VII.) If we look at things in this way, we may think of the Inquirer's intellectual predicament as setting the task of eliminating uncertainty by closing the "logical gap" between premises and conclusion.

Admittedly, the identification of the erotetic and the cognitive psychologist's positions is partial and incomplete, but once the Inquirer is allowed to make interrogative moves (i.e., ask information-seeking questions of the Oracle), the difference between the logical and cognitive situations becomes less pronounced. Nevertheless the difference between a logical and a cognitive gap does not disappear. The logical represents an ideal that may be grasped *a priori* whereas the psychological requires empirical *a posteriori*

4. There is a shift here from our previous use of this term; in Chapter II above, the intellectual predicament was viewed more as a state of the student than a task he or she must perform; but remember that the intellectual predicament defines the "epistemological ought" which it is the task of the teacher to carry out. Here, the teacher's task becomes the student's (or Inquirer's).

considerations. The similarities between the two, however, permit us to construct a logical home for cognitive psychology that helps us better understand the enterprise of cognitive research. For one thing, this similarity will allow us to reconceptualize many of the erotetic teaching moves discussed in earlier chapters as strategic options in the process of problem solving (and vice versa). Another result is that the overlap between the notions of logical gap and cognitive gap allows us to offer some measure of a problem's difficulty.[5]

The Logical Home of Problem-Solving Processes

For the most part deductive heuristics (or erotetic heuristics) along with the moves of pedagogical and dialogical strategy discussed in earlier chapters provide the logical constraints and possibilities placed upon and available to the Inquirer in the process of solving a given problem. It will only be necessary now to give this logical superstructure some cognitive and information-processing content.

Our earlier admonition that "everything else being equal, students ought to ask and have answered those questions that provide the most information regarding their intellectual predicaments" (Chapter IV), may now be given fuller meaning. As a rule any Inquirer ought to ask those questions that eliminate the most uncertainty or, equivalently, contribute the largest girder to the bridge that closes the cognitive gap presented to the Inquirer by the problem. In other words, whenever possible the Inquirer should ask the "big" question immediately. The exceptions governing

5. We shall not here demonstrate this possibility; it hinges upon counting the presuppositions in the task environment; the measure of a problem's difficulty would then be the number of presuppositions and their relations in the task environment that are missing from the problem space. Full development of this sketch would require us to go too far afield for this chapter.

this general heuristic rule have already been considered in our discussion of partial answerhood and the "chain of questions." The major block to petitioning the principle (or problem) is the necessity of first securing the requisite presuppositions in order to avoid obtaining dangerously ambiguous answers. Obtaining the necessary new presuppositions is the chief reason for asking questions that can receive partial answers.

We may now take the notion of information-ordering questions quite literally. Because of the interconnections among presuppositions (or nodes) in LTM it is possible to derive more information from LTM than was, in one sense, originally stored there (Bower 1978). Information-ordering questions structure information in such a way that the Inquirer can recognize the connections and thereby derive additional information from LTM. Recall that we described information ordering questions as eliciting information implicit in one's background knowledge. This intuitive idea is justified by information processing models, particularly in the notion of "metacognitive" information processing - the Inquirer's knowledge of his or her own knowledge (Fredericksen 1984).

Another function of such metacognitive questions is to elicit tacit knowledge. Tacit knowledge is simply knowledge already stored in LTM that must be retrieved and placed in short-term or working memory (STM). Working memory contains information that is actively being used. Failure to activate presuppositions stored in LTM would needlessly deplete the information available in the original problem space and thereby give rise to unnecessary difficulties in problem resolution. Tacit knowledge differs from implicit knowledge in that in the former the knowledge was at one time explicit and only needs to be reactivated whereas in the latter inferences must be drawn from previously existing information in order to render the implicit knowledge explicit.

New presuppositions - from whatever source - open up new lines of inquiry. One way of obtaining new presuppositions is, as previously indicated, from the answers to earlier questions.

Another way is simply to guess, that is, to formulate questions containing a *hypothetical* presupposition. Hypothetical questioning arises in the context of problems that require productive thinking. In science, this strategy manifests itself in the hypothetico-deductive methods wherein the Inquirer postulates a hypothesis, deductively draws statements describing empirical consequences from the hypotheses and other presuppositions, and then asks nature an experimental question. A favorable answer is then said to confirm the hypothesis.

There are dangers here; cognitive psychologists (e.g., Mynatt, Doherty, and Tweney 1977, and Moshman 1979) have demonstrated a dangerous "confirmatory bias" in solving problems that required hypotheses for their solution. This bias poses a serious threat to successful inquiry, since the hypotheses chosen can instantiate false presuppositions as easily as true ones.

False presuppositions may arise in many different ways - "confirmatory bias" is one, false instruction is another. There are others as well. False presuppositions need not prove fatal to inquiry, although they often are: it is possible to be right for the wrong reasons. False presuppositions do not preclude the inquiry from succeeding, as the history of science amply testifies, but they limit the number of questions that can be answered correctly and must eventually misdirect inquiry and mislead the Inquirer. Detecting and correcting such false presuppositions and the relations they enter into is not unlike "debugging" faulty problem-solving programs (Brown and Burton 1978).

Newell and Simon (1972) have found that many problem solvers spontaneously gravitate toward a heuristic they call "means-ends analysis." In this heuristic, Inquirers will repeatedly compare their present state with the desired end state and attempt to determine the size of the logical gap (our term) between where the Inquirer is and where he or she would like to be, in hopes of determining what can be done to close the gap. Means-ends inquiry is linear and proceeds sequentially with each step intended to draw the Inquirer ever closer to the solution. In the logic of questions and answers each step would correspond to asking a "smaller" information-seeking question with the goal of eliminating some of the epistemological alternatives to the actual situation, i.e., the answer to the "big" question. The logic of means-ends analysis is captured by what we called in an earlier chapter a "chain of questions."

One shortcoming of means-ends analysis is that it is usually only effective with well-structured problems and that it rarely provides for the introduction of novel concepts. Erotetic logic illuminates these shortcomings. Note that if the actual solution is not in the initial set of epistemological alternatives, means-ends analyses proceeding by smaller information-seeking questions intended to eliminate alternatives will not generate them, and so will not turn up the answer. Dialectic, as a logic of discovery, provides some help here. First, in dialectical reasoning, we may introduce new presuppositions by hypothesis. Second, answers from the Oracle may introduce new presuppositions into the inquiry. New presuppositions alter the set of epistemological alternatives compatible with the presuppositions of the Inquirer and, ideally, the actual situation may be found in the altered set. Productive thinking calls for divergent as well as convergent thinking.

Perhaps the best known problem solving heuristic consists in merely instantiating the problem statement, i.e., providing a visual or otherwise concrete image in place of an abstract linguistic statement. This is precisely the function of the geometrical diagrams drawn by Socrates in the course of teaching the slave-boy how to solve the problem of doubling a square. In the logic of questions and answers this is tantamount to instantiating some of the available presuppositions in the problem space.

Presuppositions may be instantiated so as to formulate a question. This is commonly the case when the Inquirer instantiates hypothetical presuppositions - as when the Inquirer "tries" an auxiliary construction (in geometry) or sets up an experiment (a question put to nature), usually with the hypothesis as the dependent variable. Remarkably, those who study problem solving tend to ignore this heuristic altogether. It is, however, easy to grasp why this heuristic works. By representing the stored information in external space it is no longer necessary to store it in working memory whose limited storage capacity (estimated as 7 ± 2 items) may then be freed to process other information (Miller 1956).

Conclusions

This discussion of the relations between cognitive psychology, information-processing theory and erotetic logic smacks of being nothing more than a vocabulary lesson, or a translation of one set of words into another. In a way, this is so; one thing that becomes obvious is that the "empirical" theoretical structure of those other fields is remarkably similar to the logical theory that lies behind our erotetic theory of teaching. This should not be surprising, perhaps, since high level theory building in any field combines elements of *a priori* reasoning and empirical investigation of the "fit" of such reasoning to the world.

What we have tried to show in this chapter is the power of erotetic logic to illuminate the logical structure of the background theories of cognitive psychology. Insofar as we have been successful, this should provide a fruitful direction for further research on this intersection of logic and empirical psychology. There is good reason for believing that empirical studies of teaching and learning (including problem solving as a model for pedagogical reasoning and for learning) can proceed with a firm logical base.

CHAPTER X

EROTETIC PROSPECTS

An underlying assumption of the foregoing discussion of teaching is that the Wittgensteinian doctrine that philosophy leaves everything as it is is not appropriate where practice, research and philosophy meet. What is needed at this juncture is not only clarity about a fuzzy concept, but reasons for making decisions about how to delimit the concept for use in these contexts. Philosophers must become theorists, arguing for a theoretical definition of teaching which will prove its usefulness by helping to explain what goes on in the domain under consideration by scientists and practitioners. To be adequate, such an analysis has to connect with the ordinary concept in consistent ways, but it must be rigorous and precise enough to help in the explanation of what is going on.

By way of conclusion, let us suggest some areas in which the erotetic theory of teaching shows some additional promise. Our demonstration will be more of a sketch of prospects for further analysis and research than a set of conclusions of completed analyses. Where would this sort of approach help educationists to understand and further develop the central issues in teaching and related areas? We will consider first some areas in current pedagogical research and theory, and then conclude by considering a new-but-old paradox generated by the concept of teaching.

Time on Task

One of the more interesting findings of recent research on teaching is that time spent directly on academic work correlates strongly with academic achievement, while time spent on games, non-academic activities, and individual unsupervised work correlates little or negatively. This research was aimed directly at "open-school" advocates, whose central contention was that allowing students to "mess around" would lead to better results than highly structured lessons and strict supervision of students. Time-on-task research has not supported this belief; it has led to more structured models of teaching. But as Soar and Soar (1976) point out, it is not a direct correlation that provides the support for conclusions about teaching methods - it is, rather, what they call an "inverted-U" correlation, where significant improvement is correlated with some increase in time on task, but where too much time makes the correlation fall off. But there is no explanation for these findings inherent in the correlations; the explanation must be found elsewhere.

Here is an erotetic possibility: in any area other than rote learning (where fatigue is probably the major factor), more time spent on a particular pedagogical problem is likely to involve the repetition of material. But once a question is answered for and understood by a student, more time spent merely repeats already completed tasks; reanswering the same question does not promote further achievement, however measured. Answering a question, however, opens up new possibilities for instruction; new intellectual predicaments replace those whose epistemological oughts have been satisfied. What tasks are attended to is at least as important as how much time is spent; and these tasks may most accurately be portrayed in terms of the students' intellectual predicaments. The erotetic analysis of teaching suggests a way of getting at the reasons for the importance of time-on-task and for the inverted-U correlations. Attending to the content of the instruction and to the logical flow of the interaction between teachers and pupils should be promising.

Testing the hypotheses involved in this explanation-sketch would require considerable empirical investigation, but its plausibility seems evident: the relevance of the teachers' remarks about and within the subject matter and the relevance of the content of the time spent on academic materials to the students' intellectual predicaments provides a beginning explanation of what has happened in the classrooms under investigation. Further study of this dimension of teaching, using all the ingenuity of modern investigative methods, should lead to a greater understanding of how classroom instruction can be explained and improved.[1]

Motivation

A constant problem for teachers and researchers alike is how to conceive of and study students' motivation to learn the materials under consideration. Standard approaches all too often attend to factors external to the pedagogical situation - the rewards that students will receive for learning (in the form of personal satisfaction, gold stars, praise from significant people, etc.). If we wish to comprehend why a student learns a particular bit of material, however, it should be promising to attend to the student's intellectual predicaments, the resultant epistemological oughts, and the questions and answers that make up the pedagogical relationship. To see the point of this, it is necessary to detour through some features of the explanation of behavior by the agent's motives.

In ordinary life, we use the term 'motive' exclusively in explaining voluntary actions. The question, "What was Albert's motive for doing *that*?" implies that the questioner believes that the action was done voluntarily. "What was Albert's motive for falling down," suggests that he fell down on purpose. And the Freudian

1. For more extensive discussion of the problem of time on task research, see the papers collected in Fisher and Berliner 1985.

question, "What was his motive for making that slip of the tongue?" served to bring into the realm of voluntary action, and thereby into the realm of explanation by purpose, of a wide range of otherwise inexplicable happenings.

In pedagogical contexts, concerned with explanations of a person's learning or understanding something, if we ask "What was his motive for learning that?" we do the same thing with learning that Freud did with parapraxes: we view learning as voluntary action. To ask for the motive for learning something implies that a voluntary-action explanation is relevant and proper. This is out of the ordinary, for learning is not necessarily an action. One can learn something voluntarily, but one need not do so; thus, asking for the motive for learning or understanding something begs the question of what type of explanation is appropriate.

Teacher's concerns with motivation, of course, are practical; their question is not the explanatory, "Why did Albert learn that?" but the manipulative, "How can I motivate Albert to learn this?" The shift reflects the school teacher's constant problem of arousing and maintaining student interest in the subject at hand. It takes teachers directly into psychological theories of motivation, with their concepts of "achievement motivation," need-reduction," "drives," and so forth. We probably need to use concepts and theories like these if the concern is with the students' motivation for entering into the practices and procedures of schooling or with their rejection of such things.

But if our concern is for a student's motive for learning or understanding some particular subject, we are in different territory. What is needed here is specific attention to the subject and to the student's predicament with regard to that subject. The motive is not *psychological* in the sense of being a drive or a push, but rather *epistemological*, having reference to the intellectual predicament and epistemological ought. The explanation of learning in pedagogical contexts, and any manipulative conclusions that teachers may draw from such explanations, must be couched in the epistemological language of intellectual predicaments.

An erotetic approach to motivation looks promising, then, as a way of throwing a different kind of light on the motivational problems of teaching. It throws a spotlight on an area that is too easy to overlook in a non-intentional view of the situation; it requires attention directly to individual students and their concerns about the world and the subject matter through which they might legitimately expect enlightenment. Again, there is a complex set of logical and empirical studies in the future.

Although it is not currently fashionable, some cognitive psychologists of a generation ago did focus on activities such as "exploration" or "curiosity" - with typical research efforts generally resulting from the observation that such activity often occurred in the absence of any biological need state, and thus exist simply for their own intrinsic rewards (Hunt 1960; Fowler 1965; Wildman 1974). Berlyne (1954, 1960, 1962) developed a theory of *epistemic* (knowledge related) curiosity; he argues that epistemic motivation or curiosity is the product of conceptual conflict. Festinger (1957) relevantly speaks of "cognitive dissonance." Both approaches are related to the questions that students ought to ask about their worlds. We know of nothing in the literature discrediting this line of research; it appears simply to have lost momentum. It seems to us that studies of epistemic curiosity reinforced by such notions as intellectual predicaments and epistemological oughts could go far toward reopening and revitalizing this line of investigation.

The Practical Side of Teaching

One advantage of the erotetic theory of teaching is that it brings together the researcher and practicing teacher in one language, one set of concepts.[2] How the theory is relevant for practicing teachers

2. For another approach to this issue, see Petrie 1968.

is not directly obvious, however. A brief survey of some places where question-answering is relevant to the teacher's reasoning and practical activities will be suggestive if not conclusive. We will divide these into three: planning, execution and evaluation.

Planning. Given that there is a particular subject matter on which the teacher and students focus, the pedagogical question is how to make it comprehensible and retainable for the students. When the teacher considers what the students already know, thinking of what questions they ought to be asking provides the beginning form of the lesson. If the students have presuppositions which are incorrect, the teacher's task is to work with those until the main questions are there to be asked and answered. If the presuppositions are in order, but there are gaps in the students' knowledge or understanding, the task becomes figuring out how those gaps can be filled. Dealing with different students' presuppositions complicates this, of course, for the teacher faced with more than one student, but it is no more complicated than any other attempt to get at individual differences, since the crucial individual differences in pedagogical contexts are those that relate to the questions that the students ought to ask about the focal subject matter. Trying to hit the mark in teaching is trying to answer the questions.

Execution. As in any interaction among people, much depends upon the materials at hand, the audience or auditors, the environment in which teaching takes place. If teachers are alert to the cues which indicate the presuppositions of questions the students might ask, they can meet them fairly directly; the excitement of teaching - of discovery almost - is felt by students and teachers alike when a lesson clicks. It would be interesting to find out (empirically) if this happens when teachers speak directly and relevantly to the students' epistemological oughts. No teacher should limit a lesson to a pre-planned exposition of a point without regard to the questions that arise as the lesson progresses: the teacher must be alert to the unasked question, to the look of puzzlement, to the relevant digression to meet a false presupposition, to the relationships of other ideas to the questions at hand. No one should think that this is easy, but structuring and executing lessons according to students' questions provides a different dimension to the idea of pedagogical relevance, a way of structuring practice that is not given by the usual causally determined picture of teaching.

Evaluation. There are two topics here, evaluation of student progress and evaluation of teaching itself. The relevance of teachers' questions to diagnostic testing hardly needs to be recalled, except to point out that the center of attention should be the students' intellectual predicaments and epistemological oughts with regard to the subject matter in question. Teachers should be searching for the students' present state of knowledge in order that teaching can continue; the crucial dimension is not a matter of mere quantity of knowledge, but rather an exploration of the presuppositions that students have concerning the subject under consideration, the presuppositions that give rise to questions that they ought to be asking. Similarly, achievement testing can be structured more adequately in accord with erotetic logic than with standard propositional forms of logic.

Equally important, though, is the evaluation of teaching itself. Teachers do not always succeed in their pedagogical plans. Improvement of future teaching depends heavily upon the diagnosis of failure or near-failure in teaching. There are many dimensions here, needless to say, not all of which are attributable to missed questions (a teacher's nastiness to students may be a significant reason for pedagogical failure, but it is not necessarily related to the logic of questions); but when teaching is viewed as centrally concerned with answering students' unasked questions, one thing that should be central to any diagnosis is the degree to which the teacher was actually answering questions that were "there." That is, failure in teaching might often be traced to the teacher's lack of attention to the students' real epistemological oughts.

Matters of strategy in answering are also relevant - whether it is better to make the students' questions explicit or to structure the lesson so that it makes the intellectual predicaments obvious enough for the students themselves to ask the question is open to argument and empirical investigation. Research on "adjunct questions" in written materials might be relevant here, although it seems relatively unfocussed (Wittrock and Lumsdaine 1977; Hamaker 1986).

This sketch suggests that attention to the erotetic dimension of pedagogical reasoning and practice should be fruitful in the training and practice of teachers. If teaching is viewed as essentially concerned with answering students' questions with regard to particular subjects, the attention of teachers turns to central issues of the profession. The careful teacher is trying always to promote the students' intellectual development; insofar as this development is active and questioning, the teacher's attention should be focused on the interaction of question and answer; making a lesson relevant to individual students is never easy, but attending to their intellectual predicaments paves the way for an honest and open relationship between teacher, students and subject matter. Without this, there probably is no pedagogical hope, however pleasant the interaction of teacher and student may be.

Conclusion: The Importance of Teaching

Let us close by considering an unusual claim, one that initially appears absurd. Davis (1977) has constructed an anomaly that leads to the conclusion that teaching is impossible. In the end, as we shall argue, the claim fails, but what is remarkable is that the reasons for its failure may not be apparent unless one has a theory like ours that explicitly acknowledges the intentional and dialogical properties of teaching.

Defining 'teaching' as "teacher directed or designed interactions having learning by the student as the intended outcome," Davis states, "Teaching takes place when some information (or some skill), is communicated from teacher to pupil" (1977, p. 304). Davis takes the notion of communication as his central focus, and argues that what is communicated "must be whatever information is *shared* by speaker and hearer" (p. 306, emphasis ours). "Communication which is intentional," Davis continues, "will be deemed successful if what is communicated is what the speaker intended to communicate" (p. 306). From this account of communication Davis comes up with a pedagogical version of Meno's paradox:

> If teaching is to be of use in the cases for which it is needed, it must be possible in cases in which the assumption of shared systems of beliefs fails. But is this possible? For something to be taught it must be communicated. For it to be taught it must be a consequence of the material presented by both the teacher's system of beliefs and the pupil's system of beliefs. Insofar as the pupil's and the teacher's beliefs do not coincide, communication does not take place. But insofar as the pupil's and the teacher's beliefs do coincide, teaching is unnecessary. So, insofar as teaching can occur it is unnecessary. And, insofar as teaching is necessary, it is impossible. (p. 307)

Davis gets out of this bind by proposing a distinction between normal and non-normal communication, where the latter does not require a set of shared beliefs but rather a recognition that a particular communication is to be taken as coming or leading toward a logical consequence of a belief or set of beliefs of either the speaker or auditor. Drawing on Grice (1969), Davis constructs his own example of what he means by non-normal communication.

> If someone, whom I know to believe that bulls are gentle, tells me that John is as gentle as a bull, then I can take him as saying that John is gentle and he will have communicated this

> to me, despite the fact that I don't believe that bulls are gentle. (p. 307)

Non-normal communication is a rare form of communication, one that calls for "special knowledge" about the perhaps mistaken beliefs of others. Since, according to Davis, "the pupil's system of beliefs is infinite, interdependent and private," and since "The hypothetico-deductive method is the only reliable device we have for attaining knowledge about what is beyond the limits of our observation," the teacher, in engaging in non-normal communication with students, depends upon his or her hypotheses about the students' belief systems; communication is seen as possible only when the student's belief system is a "sub-set of the teacher's belief system" (p. 309). But this situation is considered a rarity; when assumed by the teacher, it is highly hypothetical and tentative, if not unverifiable. Davis ends with a pessimistic conclusion:

> But we haven't sufficient grounds for assuming the pupil's system of beliefs is a subset of the teacher's system of beliefs. If teachers were to teach pupils of basically their own social status in a monolithic and conservative society, we could reasonably hope this assumption to hold. We know, however, that present North American societies are not societies of this sort.
>
> We seem to have arrived at another, and a final, dead end. We are left with a problem that is essential to teaching - it arises from the concept itself. We cannot solve the problem, and its existence ensures that teaching is not possible except in a quite restricted special case. This is an uncomfortable conclusion, but I see no way to avoid it. I see only one way in which its discomforting effects can be evaded - give up teaching. This alternative is drastic, but is not impossible. Teaching isn't the only device available for human education, it is just one we happen to have been using rather extensive-

ly recently. Can we, then, stop attempting the impossible, and instead look seriously at some of these other alternatives? (p. 309)

The reader should share our discomfort at Davis's conclusion: something seems wrong in any argument leading to the conclusion that teaching is impossible. But if, when fully fleshed out, Davis's argument is valid, then one or more of his premises regarding communication, belief and the epistemological determination of belief must be false - or else we are driven to his conclusion.

More disturbing, perhaps, is the helplessness of non-intentional versions of teaching in dealing with arguments like Davis's. The reason for this is obvious: what could they say? There is nothing in the vocabulary of, for example, the process-product research tradition that enables it to deal with the intentionally loaded paradox that Davis constructs. The technical language of numerically quantified variables, correlations, and so forth, cannot handle such arguments. The language of communication (or alternatively, information processing or belief) is appropriate for higher level cognitive functions; and that language, as we hope we have shown, is essential in any discussion of teaching.

The conceptual impoverishment of a strictly-conceived process-product tradition of research on teaching, as we have shown above, appears very early within the tradition itself - as, for instance, when Gage (1978) turns studies of fairly low level correlations among at most three or four variables over to artist/practitioners who are then expected to link them up with larger numbers of variables, control for interaction effects and contextualize them in communication with students. These "artists" are expected to convert research into practice without guidance from the theory behind the research. To add stress to the situation, the artists may expect to be held accountable by evaluators for their performance (Macmillan and Pendlebury, 1985; Shulman 1987). But it is at precisely this point that teachers want and need help;

it is also precisely at this point that most current research abandons them. They are left to fend for themselves against administrators, evaluators and philosophers like Davis.

The problem is serious, both practically and theoretically. If a research tradition is unable to respond to arguments like Davis's, the research findings it produces are rendered superfluous since it is unable to show even how teaching is possible, much less how research findings are to be organized so as to make *effective* teaching possible.

The situation looks better for an approach like erotetic theory of teaching, since it is able to handle the problems of intentionality in teaching more adequately. Davis's conclusion overlooks the possibility of a different conception of communication between teacher and student. When teaching is viewed as a question-answering activity, the possibility of communication opens up entirely - for the teacher must be seen as talking directly to the students' belief systems, to intellectual predicaments in which the students' questions are founded. The criteria of satisfaction for questions at least assume logical contact, even if they cannot guarantee communication.

The concerns of communication are explicitly addressed by pedagogical and dialogical strategy as outlined above. Indeed, the point of the dialogical teaching game is the exchange of information, or communication.

Davis takes the hypothetical nature of determining the students' belief systems as being a problem for teachers. If the problem is merely whether the students' beliefs are a subset of the teacher's, then it is a problem indeed, for teaching seems to be called for only when the students do not already have the beliefs that the teacher is trying to communicate or justify. But if it is that determining anyone's belief system is hypothetical, it fades into a semi-technical problem of determining the best methods of supporting particular hypotheses about those beliefs. There is considerable discussion in the philosophical literature on just this topic, and the best of the test-and-measurement theories are also relevant. Even here, of course, the logic of questions is crucial; for what is to be discovered is the state of each student's knowledge and this is done in part through questioning, in part through inferences based upon observation of the student's non-academic behavior, and so forth. This is difficult, but not particularly mysterious. In fact, Davis is just wrong to say that the hypothetico-deductive method is the only reliable device we have for attaining knowledge of unobservables. Plato, for instance, originally developed his dialectic in order to engage in inquiry about the unperceivable forms. Dialectic reasons hypothetically, but not necessarily hypothetico-deductively. More to the point, teachers' diagnostic questions and inferences to the presuppositions of students' explicit questions are two ways of getting at students' beliefs fairly directly through discourse.

In the end, Davis's paradox is no paradox; his pessimistic conclusion about teaching is, therefore, uncalled for.

Furthermore, Davis's recommendation that other methods of education replace teaching overlooks the moral values that are involved in teaching when it is conceived as something other than mere passing of ideas from one person to another. When teaching is seen as an important form of human interaction which includes the respect shown by a teacher who would answer the questions implicit in students' intellectual states in a responsible way, the importance of teaching - and of the decision to *teach* particular subjects to our children - should come clear. To leave to chance or to some less open form of bringing about learning (perhaps by manipulation of the environment or of persons through conditioning or indoctrination) is to abdicate our own respect for truth and our love of our children. The decision to teach, to engage in this kind of interaction with others, puts a unique import upon the subjects learned that is lost when other routes to learning are taken.

Teaching is a significant social option.[3] It is only one of many possible ways in which we may educate our children and ourselves. When it is carried out in a spirit assumed by the erotetic conception of teaching - in which the questions answered are those of the student - attention is drawn to the finest values of human interaction: truth, honesty, and openness.

We have claimed that the erotetic conception of teaching is neutral with regard to the subjects studied. But it should be clear that it is not neutral with regard to the ways in which *any* subject should be studied. Its emphasis is upon reason, upon truth, and upon honesty. An attempt to answer students' questions that is guided by such values cannot be all wrong.

But our analysis of teaching will undoubtedly leave some readers uncomfortable; surely, it will be thought, teaching is much more than merely answering unasked questions (or even asked questions). The teacher acts as a model, a guide, a motivator, an exemplar. The teacher must show students where errors are made, drill them for the development of skills and habits, in short do a great deal more than the minimum suggested by this analysis. Read any book on teaching, any discussion of how to put together a lesson, of what elements are involved in teaching others anything, and the discussion will be rich with metaphor, deep in psychoanalytic concepts or their equivalents, and fraught with moral dilemmas. (See, for example, Passmore 1980; Barzun 1944; Joyce and Weil 1980.) Examine Miss Brodie's effects on the Brodie set (Spark 1961), see how John Updike's teachers delve through layers of sexuality and *angst* (Updike 1959) - teaching is much richer than any theory like ours can catch.

And of course that is true - for the theory we have tried to develop here is not designed to provide a picture of everything that happens in teaching or in schools. No one could run a school

3. Scheffler (1960) emphasizes this point strongly. It should be clear that our analysis of teaching is related to the "restrictions of manner" that lie at the center of his discussion.

or even plan a complete lesson using only the notions of epistemological oughts and intellectual predicaments. Answering one student's unasked questions may be very different from answering another's, even if the questions have the same form - for the epistemological oughts find their homes in complex patterns of belief and background, within even more complex patterns of motivation and ways of behaving. Furthermore, different teachers have different styles of personal interaction within which they approach the problems of tactics and strategy in answering their students' questions.

We do claim that the erotetic concept - and theory - of teaching gives a much tighter analysis of the "core" of teaching - the *sine qua non* - than has previously been given; in addition, it shows unexpected ways in which many other features of teaching fit around the core. What we have tried to get at is the equivalent of the defining conditions of games like bridge and golf - the activities which constitute the games or activities, within whose context the players or agents determine strategies, tactics, and procedures, and around which institutions like bridge parties and country clubs are built. Just as golf is more pleasantly played in comfortable or traditional surroundings than in a rocky field, teaching can probably be more adequately carried out in optimum situations than on the end of a log. But the surroundings are not to be confused with the game or with teaching; the latter justify the former rather than vice versa. Research which attends to the surroundings of the core of teaching has not been useless, but its importance has probably been misunderstood.

It is our belief that the type of analysis we have given here will enable both the science and practice of teaching to be better understood - and that the science and the practice will be more closely related because they use the same language and concepts. Whether this type of activity can be successfully institutionalized is a separate question, but one that is worthy of an answer. For teaching is too important to leave in its present state of confusion.

REFERENCES

Anderson, J. R. 1981.'Concepts, Propositions, and Schemata: What are the Cognitive Units?' In H. E. Howe & J. H. Flowers (eds.), *Nebraska Symposium on Motivation, 1980: Cognitive Processes*. Lincoln: University of Nebraska Press.

Anderson, R. C. 1984. 'Some Reflections on the Acquisition of Knowledge.' *Educational Researcher* 13 (9):5-10.

Anscombe, G. E. M. 1965. 'The Intentionality of Sensation: A Grammatical Feature.' In J. R. Butler, ed. *Analytical Philosophy*, Second Series. Oxford: Basil Blackwell.

Aristotle. *Posterior Analytics*. In R. McKeon, ed. 1941. *The Basic Works of Aristotle*. New York: Random House:110-186.

Austin, J. L. 1962. *How to do Things with Words*. Oxford: Oxford University Press.

Ausubel, D. P. 1968. *Educational Psychology: A Cognitive View*. New York: Holt, Rinehart, and Winston.

Barzun, J. 1944. *Teacher in America*. Boston: Little Brown.

Berliner, D. C. 1984. 'Making the Right Changes in Preservice Teacher Education.' *Phi Delta Kappan* 66:94-96.

Berlyne, D. W. 1954. 'A Theory of Human Curiosity.' *British Journal of Psychology* 45:180-191.

Berlyne, D. W. 1960. *Conflict, Arousal, and Curiosity*. New York: McGraw-Hill.

Berlyne, D. W. 1962. 'Uncertainty and Epistemic Curiosity.' *British Journal of Psychology* 53:27-34.

Beta-Blocker Heart Attack Trial Research Group. 1982. 'A Randomized Trial of Propanolol in Patients with Acute Myocardial Infraction.' *Journal of the American Medical Association* 247:1707-1713.

Bloom, B. S. 1980. 'The New Direction in EducationalResearch: Alterable Variables.' *'Phi Delta Kappan* 61:382-385.

Bower, G. H. 1978. 'Contacts of Cognitive Psychology with Social Learning Theory.' *Cognitive Therapy and Research* 2:123-146.

Bromberger, S. 1965. 'An Approach to Explanation.' In R. J. Butler, ed. *Analytical Philosophy*, Second Series, Oxford: Blackwell.

Broudy, H. S., Ennis, R. H. and Krimerman, L. I. (Eds.) 1973. *Philosophy of Educational Research.* New York: Wiley.

Brown, H. I. 1977. *Perception, Theory and Commitment: The New Philosophy of Science.* Chicago: University of Chicago Press.

Brown, J. S. and R. R. Burton. 1978. 'Diagnostic Models for Procedural Bugs in Basic Mathematical Skills.' *Cognitive Science* 2:155-192.

Bruner, J. S. 1960. *The Process of Education.* Cambridge: Harvard University Press.

Cronbach, L. J. 1975. 'Beyond the Two Disciplines of Scientific Psychology.' *American Psychologist* 30 (2):116-127.

Davidson, D. 1963. 'Actions, Reasons, and Causes.' *Journal of Philosophy* 60. Reprinted in Davidson, 1980:3-20.

Davidson, D. 1980. *Essays on Actions and Events.* Oxford: Oxford University Press.

Davidson, D. 1984. *Inquiries into Truth and Interpretation.* Oxford: Oxford University Press.

Davis, B. 1977. 'Why Teaching Isn't Possible.' *Educational Theory* 27:304-309.

Dewey, J. 1916. *Democracy and Education.* New York: Macmillan.

Dewey, J. 1933. *How We Think.* New York: Heath.

Dietl, P. J. 1973. 'Teaching, Learning and Knowing.' *Educational Philosophy and Theory* 5:1-25.

Dillon, J. T. 1982. 'Superanalysis.' *Evaluation News.* November 1982:35-42.

Dillon, J. T. (ed.). 1988. *Questioning and Discussion: A Multidisciplinary Study.* Norwood, NJ: Ablex. [Forthcoming].

Doyle, W. 1978. 'Paradigms for Research on Teacher Effectiveness.' *Review of Research in Education* 5, Itasca, IL: Peacock.

Duhem, P. M. 1906. *The Aim and Structure of Physical Theory.* Paris. Trans. P. P. Wiener. 1956. Princeton: Princeton University Press.

Ennis, R. H. 1973. 'On Causality.' *Educational Researcher* 2 (6):4-11.

Ennis, R. H. 1982. 'Abandon Causality?' *Educational Researcher* 11 (7):25.

Ennis, R. H. 1986. 'Is Answering Questions Teaching?' *Educational Theory* 36:343-347.

Ericson, D. P., and F. S. Ellett, Jr. 1982. 'Interpretation, Understanding and Educational Research.' *Teachers College Record* 83:497-513.

Ericson, D. P., and F. S. Ellett, Jr. 1987. 'Teacher Accountability and the Causal Theory of Teaching.' *Educational Theory* 37:277-293.

Feinleib, M., C. Lenfant, and S. A. Miller. 1984. 'Letter: Hypertension and Calcium.' *Science* 226:384, 386.

Fenstermacher, G. D. 1979. 'A Philosophical Consideration of Recent Research on Teacher Effectiveness.' in L. S. Shulman, ed., *Review of Research in Education 6.* Itasca, IL: F. E. Peacock.

Festinger, L. 1957. *A Theory of Cognitive Dissonance.* Stanford: Stanford University Press.

Fisher, C. W., and D. C. Berliner, eds. 1985. *Perspectives on Instructional Time.* New York: Longman.

Fodor, J. A. 1981. *Representations: Philosophical Essays on the Foundations of Cognitive Science.* Cambridge: MIT Press.

Fowler, H. 1965. *Curiosity and Exploratory Behavior.* New York: Macmillan.

Frederiksen, N. 1984. 'Implications of Cognitive Theory for Instruction in Problem Solving.' *Review of Educational Research* 54:363-407.

Gage, N. L. 1963a. 'Paradigms for Research on Teaching.' In N. L. Gage, ed. *Handbook of Research on Teaching*. Chicago: Rand-McNally.

Gage, N. L. 1963b. 'Preface.' In N. L. Gage, ed. *Handbook of Research on Teaching*. Chicago: Rand-McNally.

Gage, N. L. 1966. 'Research on Cognitive Aspects of Teaching.' In *The Way Teaching Is*. Washington: Association for Supervision and Curriculum Development and NEA Center for the Study of Instruction.

Gage, N. L. 1978. *The Scientific Basis of the Art of Teaching*. New York: Teachers College Press.

Gage, N. L. 1983. 'When Does Research on Teaching Yield Implications for Practice?' *The Elementary School Journal* 83:492-496.

Gage, N. L. 1984, 'What Do We Know about Teaching Effectiveness?' *Phi Delta Kappan* 66:87-93.

Garrison, J. W. 1985. 'Dewey and the Empirical Unity of Opposites.' *Transactions of the Charles S. Peirce Society*.

Garrison, J. W. 1986. 'Some Principles of PostPositivistic Philosophy of Science.' *Educational Researcher* 15 (9):12-18.

Garrison, J. W. 1987. 'The Impossibility of Atheoretical Educational Research.' *Journal of Educational Thought*. [Forthcoming].

Garrison, J. W. and C. J. B. Macmillan. 1984. 'A Philosophical Critique of Process-Product Research on Teaching.' *Educational Theory* 34:255-274.

Garrison, J. W. and C. J. B. Macmillan. 1987. 'Educational Research to Pedagogical Practice: A Plea for Theory.' *Journal of Research and Development in Education* 20:38-43.

Gasking, D. 1955. 'Causation and Recipes.' *Mind* 64:479-487.

Gergen, K. J. 1973. 'Social Psychology as History.' *Journal of Personality and Social Psychology* 26:309-320.

Glass, G. V. 1976. 'Primary, Secondary and Meta-analysis of Research.' *Educational Researcher* 5:3-8.

Glass, G. V. 1978. 'Integrating Findings: The Meta-analysis of Research.' *Review of Research in Education 5*. Itasca, IL: Peacock.

Glass, G. V., B. McGaw, and M. L. Smith. 1981. *Meta-analysis in Social Research*. Beverly Hills: Sage.

Goody, E. N. (ed.). 1978. *Questions and Politeness: Strategies in Social Interaction*. Cambridge: Cambridge University Press.

Green, J. L.1983. 'Research on Teaching as a Linguistic Process: A State of the Art.' *Review of Research in Education 9*. Washington: American Educational Research Association.

Green, T. F. 1968. 'A Topology of the Teaching Concept.' In C. J. B. Macmillan and Thomas W. Nelson, eds. *Concepts of Teaching: Philosophical Essays*. Chicago: Rand-McNally.

Green, T. F. 1971. *The Activities of Teaching*. New York: McGraw-Hill.

Grice, H. P. 1969. 'Utterer's Meaning and Intention.' *Philosophical Review* 78:162-170.

Grice, H. P. 1975. 'Logic and Conversation.' In D. Davidson and G. Harman, eds. *The Logic of Grammar*. Encino, CA: Dickenson.

Guthrie, W. K. C. 1956. *Plato: Protagoras and Meno*. Baltimore: Penguin.

Hamaker, C. 1986. 'The Effects of Adjunct Questions on Prose Learning.' *Review of Educational Research*. 56:212-242.

Hanson, N. R. 1958. *Patterns of Discovery*. New York: Cambridge University Press.

Heath, P. 1974. *The Philosopher's Alice*. New York: St. Martin's Press.

Hintikka, J. 1976. 'The Semantics of Questions and the Questions of Semantics.' *Acta Philosophica Fennica 28* (4).

Hintikka, J. 1982. 'A Dialogical Model of Teaching.' *Synthese* 51:39-60.

Hintikka, J. and M. Hintikka. 1982. 'Sherlock Holmes Confronts Modern Logic: Towards a Theory of Information-Seeking

Through Questioning.' In E. J. Barth and J. Martens, eds. *Theory of Argumentation*.. Amsterdam: Benjamins.

Hintikka, J., and Renes, 1974. *The Method of Analysis*. Dordrecht: Reidel.

Hirst, P. H. 1971. 'What is Teaching?' *Journal of Curriculum Studies* 3:5-18.

Hirst, P. H. 1979. 'The Logical and Psychological Aspects of Teaching a Subject.' In R. S. Peters, ed. *The Concept of Education*. London: Routledge and Kegan Paul.

Hospers, J. 1946. 'On Explanation.' *Journal of Philosophy* 43:337-356.

Howard, V. A. 1982. *Artistry: The Work of Artists*. Indianapolis: Hackett.

Hume, D. 1739, 1740. *A Treatise of Human Nature*. Charles W. Hendel, ed. 1955. Indianapolis: Bobbs-Merrill.

Hume, D. 1777. *An Enquiry Concerning Human Understanding*. Eric Steinberg, ed. 1977. Indianapolis: Hackett.

Hunt, J. McV. 1960. 'Experience and the Development of Motivation: Some Misinterpretations.' *Child Development* 31:489-504.

Joncich, G. M., ed. 1962. *Psychology and the Science of Education: Selected Writings of Edward L. Thorndike*. New York: Teachers College Press.

Joncich, G. M. 1968. *The Sane Positivist: A Biography of Edward L. Thorndike*. Middletown, Conn.: Wesleyan University Press.

Joyce, B. and Weil, M. 1980. *Models of Teaching*. Englewood Cliffs, NJ: Prentice-Hall.

Kennedy, M. M. 1984. 'How Evidence Alters Understanding and Decisions.' *Educational Evaluation and Policy Analysis* 6:207-226.

Kenny, A. 1963. *Action, Emotion and Will*. London: Routledge & Kegan Paul.

Kintsch, W. 1972. 'Notes on the Structure of Semantic Memory.' In E. Tulving and W. Donaldson, eds. *Organization of Memory*. New York: Academic Press.

Kline, M. 1966. 'Intellectuals and the Schools: A Case History.' *Harvard Educational Review* 36:505-511.

Kolata, G. 1984. 'Does a Lack of Calcium Cause Hypertension?' *Science* 226:705-706.

Komisar, B. P. 1967. 'More on the Concept of Learning.' In B. P. Komisar and C. J. B. Macmillan, eds. *Psychological Concepts in Education*. Chicago: Rand-McNally.

Komisar, B. P. 1968. 'Teaching: Act and Enterprise.' In C. J. B. Macmillan and T. W. Nelson, eds. *Concepts of Teaching: Philosophical Essays*. Chicago: Rand-McNally.

Komisar, B. P., and Associates. 1976. *Natural Teaching Encounters*. Philadelphia: KDC Enterprises.

Kuhn, T. 1970. *The Structure of Scientific Revolutions*. Second Edition. Chicago: University of Chicago Press.

Lakatos, I. 1970. 'Falsification and the Methodology of Scientific Research Programmes.' In I. Lakatos and A. Musgrave, eds. *Criticism and the Growth of Knowledge*. Cambridge: Cambridge University Press.

Laudan, L. 1977. *Progress and its Problems*. Berkeley: University of California Press.

MacIntyre, A. 1967. 'The Idea of a Social Science.' *Proceedings of the Aristotelian Society*, Supplementary Volume.

Macmillan, C. J. B. 1968. 'Questions and the Concept of Motivation.' In G. L. Newsome, ed. *Philosophy of Education 1968*. Edwardsville, IL: Philosophy of Education Society.

Macmillan, C. J. B. 1988. 'An Erotetic Analysis of Teaching.' In James T. Dillon, ed. *Questioning and Discussion: A Multidisciplinary Study*. Norwood, NJ: Ablex. [Forthcoming].

Macmillan, C. J. B., and J. W. Garrison. 1983. 'An Erotetic Concept of Teaching.' *Educational Theory* 33:157-166.

Macmillan, C. J. B., and J. W. Garrison. 1986. 'Erotetics Revisited.' *Educational Theory* 36:355-361.
Macmillan, C. J. B., and J. W. Garrison. 1987. 'Erotetics and Accountability.' *Educational Theory* 37:295-300.
Macmillan, C. J. B. and McClellan, J. E. 1968. 'Can and Should Means-Ends Reasoning be used in Teaching?' In C. J. B. Macmillan and T. W. Nelson (eds.), *Concepts of Teaching: Philosophical Essays.* Chicago: Rand-McNally.
Macmillan, C. J. B., and S. Pendlebury. 1985. 'The Florida Performance Measurement System: A Consideration.' *Teachers College Record* 87:67-78.
Martin, J. R. 1970. *Explaining, Understanding and Teaching,* New York: McGraw-Hill.
McCarron, D. A., C. D. Morris, H. J. Henry, and J. L. Shanton. 1984. 'Blood Pressure and Nutrient Intake in the United States.' *Science* 224:1392-1398.
McKeachie, W. J. 1974. 'The Decline and Fall of the Laws of Learning.' *Educational Researcher* 3 (3):7-11.
McClellan, J. E. 1976. *Philosophy of Education.* Englewood Cliffs, NJ: Prentice-Hall.
Miller, G. A. 1956. 'The Magical Number Seven, Plus or Minus Two: Some Limits on our Capacity for Processing Information.' *Psychological Review* 63:81-97.
Morick, H., ed. 1980. *Challenges to Empiricism.* Indianapolis: Hackett.
Moshman, D. 1979.'Development of Formal Hypothesis-Testing Ability.' *Developmental Psychology* 15:104-112.
Mynatt, C. R., M. E. Doherty, and R. D. Tweney. 1977. 'Confirmation Bias in a Simulated Research Environment: An Experimental Study of Scientific Inference.' *Quarterly Journal of Experimental Psychology* 29:85-95.
Newell, A., and H. A. Simon. 1972. *Human Problem Solving.* Englewood Cliffs: Prentice-Hall.

Passmore, J. 1980. *Philosophy of Teaching*. Cambridge: Harvard University Press.

Pendlebury, S. 1986. 'Teaching: Response and Responsibility.' *Educational Theory* 36:349-354.

Petrie, H. 1968. 'Why has Learning Theory Failed to Teach Us How to Learn?' In G. L. Newsome, ed. *Philosophy of Education, 1968*. Edwardsville, IL: Philosophy of Education Society.

Piaget, J. 1971. *Psychology and Epistemology*. Translated by A. Rosin. New York: Grossman.

Plato. *Meno*. Translated by W. K. C. Guthrie. 1956. *Plato: Protagoras and Meno*. Baltimore: Penguin.

Plato. *Republic*. Translated by G. M. A. Grube. 1974. Indianapolis: Hackett.

Puff, C. R., ed. 1979. *Memory Organization and Structure*. New York: Academic Press.

Quine, W. V. 1951. 'Two Dogmas of Empiricism.' *The Philosophical Review* 60. Reprinted in Quine, 1953, and Morick 1980.

Quine, W. V. 1953. *From a Logical Point of View*. Cambridge: Harvard University Press.

Quine, W. V. 1966. 'Quantifiers and Propositional Attitudes.' In *Ways of Paradox*. New York: Random House.

Resnick, L. B. 1981. 'Instructional Psychology.' *Annual Review of Psychology* 32:659-704.

Rist, R. C. 1977. 'On the Relations among Research Paradigms: From Disdain to Detente.' *Anthropology & Education Quarterly* 8:42-49.

Rosenshine, B. 1976. 'Classroom Instruction.' In N. L. Gage, ed. *The Psychology of Teaching Methods: The 75th Yearbook of the National Society for the Study of Education.* Chicago: University of Chicago Press.

Ryan, A. 1970. *The Philosophy of the Social Sciences*. London: Macmillan.

Ryle, G. 1949. *The Concept of Mind.* London: Hutchinson's University Library.

Scheffler, I. 1960. *The Language of Education*. Springfield, IL: Charles C. Thomas.

Scheffler, I. 1965. *Conditions of Knowledge: An Introduction to Epistemology and Education*. Chicago: University of Chicago Press.

Schneider, W. and R. M. Shiffrin. 1977. 'Controlled and Automatic Human Information Processing: I. Detection, Search, and Attention.' *Psychological Review* 84:1-66.

Searle, J. R. 1969. *Speech Acts*. Cambridge: Cambridge University Press.

Searle, J. R. 1983. *Intentionality: An Essay in the Philosophy of Mind*. Cambridge: Cambridge University Press.

Shulman, L. S. 1987. 'Knowledge and Teaching: Foundations of the New Reform.' *Harvard Educational Review* 57:1-22.

Smith, B. O. 1961. 'A Concept of Teaching.' In B. O. Smith and R. H. Ennis, eds. *Language and Concepts in Education*. Chicago: Rand-McNally.

Smith, J. K. 1983. 'Quantitative versus Qualitative Research: An Attempt to Clarify the Issue.' *Educational Researcher* 12 (3):6-13.

Soar, R. S., and R. M. Soar. 1976. 'An Attempt to Identify Measures of Teacher Effectiveness from Four Studies.' *Journal of Teacher Education* 27:261-267.

Soddy, F. 1932. *The Interpretation of the Atom*. London: Murray.

Spark, M. 1961. *The Prime of Miss Jean Brodie*. New York: The New American Library. (Paperback edition published 1984.)

Strike, K. A. 1979. 'An Epistemology of Practical Research.' *Educational Researcher* 8 (1): 10-16.

Thomas, L. G. (ed.). 1972. *Philosophical Redirection of Educational Research: The Seventy-First Yearbook of the National Society for the Study of Education*. Part I. Chicago: University of Chicago Press.

Thorndike, E. L. 1906. *The Principles of Teaching*. New York: A. G. Seiler.

Thorndike, E. L. 1912. *Education: A First Book*. New York: Macmillan.

Thorndike, E. L. 1918. *The Measurement of Educational Products: Seventeenth Yearbook of the National Society for the Study of Education*, Part II. Bloomington, ILL: Public School Publishing Company. Reprinted in Broudy, H. S., Ennis, R. H. and Krimerman, L. I. (Eds.) 1973. *Philosophy of Educational Research*. New York: Wiley, pp. 19-24.

Tom, A. R. 1980. 'The Reform of Teacher Education Through Research: A Futile Quest.' *Teachers College Record* 82:15-29.

Tom, A. R. 1984. *Teaching as a Moral Craft*. New York: Longman.

Tom, A. R. 1985. 'Rethinking the Relationship between Research and Practice.' *Teaching and Teacher Education* 1.

Updike, J. 1959. 'Tomorrow and Tomorrow and So Forth.' In J. Updike, *The Same Door*. New York: Alfred A. Knopf.

Vendler, Z. 1957. 'Verbs and Times.' *The Philosophical Review* 66. Chapter 4 of Z. Vendler. 1967. *Linguistics in Philosophy*. Ithaca: Cornell University Press. (Page references are to the latter edition.)

Viehover, K. 1976. 'Some Issues in Social Science Methodology, With Suggested Solutions.' Unpublished manuscript. Washington: The Author. Cited in Gage (1978), 82-83.

Wildman, T. M. 1974. 'A Review of Epistemic Curiosity-Theory and Related Research.' Unpublished manuscript. Blacksburg, VA: The Author.

Wilson, D. 1976. *In Search of Penicillin*. New York: Alfred A. Knopf.

Wittrock, M. C., and A. A. Lumsdaine. 1977. 'Instructional Pschology.' *Annual Review of Psychology* 28:417-459.

INDEX OF NAMES

INDEX OF TOPICS